#3

8/25

MATTHEW MANDL

UNUSUAL MATHEMATICAL PUZZLES, TRICKS, AND ODDITIES

Prentice-Hall, Inc., Englewood Cliffs, NJ 07632

Library of Congress Cataloging in Publication Data

Mandl, Matthew. (date)
   Unusual mathematical puzzles, tricks, and oddities.

  Includes index.
  1. Mathematical recreations. I. Title.
QA95.M34 1984   793.7'4   84-2106
ISBN 0-13-938150-3 (case ed.)
ISBN 0-13-938101-5 (pbk.)

*Editorial/production supervision and
  interior design:* Barbara H. Palumbo
Cover design: Photo Plus Art
Designer: Celine Brandes
Manufacturing buyer: Tony Caruso

To CHARLES JOHN NICHOLSON for his talents in puzzle solving

© 1984 by Prentice-Hall, Inc., Englewood Cliffs, New Jersey 07632

*All rights reserved. No part of this book may be
reproduced, in any form or by any means,
without permission in writing from the publisher.*

Printed in the United States of America

10 9 8 7 6 5 4 3 2 1

ISBN 0-13-938150-3
ISBN 0-13-938101-5 {PBK.}

PRENTICE-HALL INTERNATIONAL, INC., *London*
PRENTICE-HALL OF AUSTRALIA PTY. LIMITED, *Sydney*
EDITORA PRENTICE-HALL DO BRASIL, LTDA., *Rio de Janeiro*
PRENTICE-HALL CANADA INC., *Toronto*
PRENTICE-HALL OF INDIA PRIVATE LIMITED, *New Delhi*
PRENTICE-HALL OF JAPAN, INC., *Tokyo*
PRENTICE-HALL OF SOUTHEAST ASIA PTE. LTD., *Singapore*
WHITEHALL BOOKS LIMITED, *Wellington, New Zealand*

CONTENTS

PREFACE — vi

1 MATHEMATICAL PUZZLES — 1
    1-1. An Exercise in Logic *3*
    1-2. The Two-Coin Puzzle *4*
    1-3. The Simple Equation *4*
    1-4. The Bountiful Choice *5*
    1-5. Trick Question No. 1 *5*
    1-6. A Factorial Question *5*
    1-7. Relatively Speaking *5*
    1-8. Double Trouble *5*
    1-9. What's Wrong Here? *5*
    1-10. Why Always 26262626? *6*
    1-11. The Simple Cube Root *6*
    1-12. What's the Cost? *6*
    1-13. The Squares Have It *6*
    1-14. Trick Question No. 2 *7*
    1-15. It's Almost Impossible *7*
    1-16. Brain Twister *7*
    1-17. What Pattern? *7*
    1-18. Upside-Down Messages *8*
    1-19. Letters in Numbers *8*
    1-20. It Sounds Complicated *9*

2 PUZZLES IN THE GEOMETRIC — 11
    Introduction *13*
    2-1. Reassemble and Gain *13*
    2-2. Paper Can't Be Stretched *14*
    2-3. The Paper Cube Blowout *15*

2-4. The Mobius Ring *18*
2-5. Wheels within Wheels *19*
2-6. The Pretzel Caper *20*

3 MATHEMATICAL TRICKS   23
Introduction *25*
3-1. Early-Solution Trick *25*
3-2. Secret Number Revealed *28*
3-3. Fast Answer Trick *30*
3-4. The Mystery Answer *31*
3-5. Another Quick Answer Trick *31*
3-6. The Hidden Product *32*
3-7. The Numerical Shell Game *33*
3-8. How Is It Done? *34*
3-9. Calculators Don't Lie *35*
3-10. This One's Easy *35*
3-11. Trick Game Is in Your Favor *35*
3-12. For You It's Easy *36*
3-13. Round-about Turn-around *36*
3-14. Two Tricky Tricks *37*

4 ODDITIES IN MATHEMATICS   41
Introduction *43*
4-1. Recurring Sequence of Ones *43*
4-2. The Original Number Reappears *44*
4-3. The Strange Oddity of Nines *44*
4-4. Same Answer Repeated *46*
4-5. The Peculiarities of Repeated Digits *47*
4-6. The Odd Alternates *49*
4-7. The Magic of 37037 *49*
4-8. Let's Look at Quotients *51*
4-9. The Digit's Sum Equals the Cube Root *52*
4-10. Why Always 22332233? *52*
4-11. The Numeral 8 Is Strange *53*
4-12. Duplicates Are Repetitive *54*
4-13. Folding Limit Is Seven *54*
4-14. Odd, Strange, and Complicated *54*
4-15. It Becomes Its Own $1/x$ *55*
4-16. It Proves Nothing *55*
4-17. Repetitive Reciprocals *56*
4-18. The 9s Stand on Their Heads *56*
4-19. Reverse and Subtract *57*

5 THE MAGIC NUMBER SEQUENCE OF FIBONACCI AND INFINITY — 59

- 5-1. The Basic Sequence *61*
- 5-2. Product Peculiarities *62*
- 5-3. The Magic Multiplier *63*
- 5-4. The Mystic Spirals *64*
- 5-5. The Golden Ratio *65*
- 5-6. The Infinity Contradiction *67*

6 ODDITIES OF COMPUTER LOGIC AND PROGRAMMING — 71

Introduction *73*
- 6-1. Logic and Math *73*
- 6-2. The Peculiar Binary System *74*
- 6-3. It's a Real Complement *80*
- 6-4. The Dialog Compromise *82*
- 6-5. The Basic System *84*
- 6-6. Some Logic System Aspects *86*
- 6-7. A Taste of Boolean Algebra *89*
- 6-8. Venn Diagrams Are Useful *90*

7 BASICS OF COMPUTER LANGUAGES AND PROGRAM STRUCTURES — 95

Introduction *97*
- 7-1. Some Fundamental Program Types *98*
- 7-2. Miscellaneous Program Languages *99*
- 7-3. BASIC Computer Language *100*
- 7-4. Flow-chart Structures *101*
- 7-5. Program Charting and Examples *103*

8 OFFBEAT PROGRAMMING EXCURSIONS — 107

Introduction *109*
- 8-1. A Bigger Piece of Pi *109*
- 8-2. Computer Has a Trick *111*
- 8-3. The Thirteen Trick Program *113*
- 8-4. Computer Trick No. 2 *114*
- 8-5. The Shell Game *116*
- 8-6. Electronic Safe Combination *117*

SOLUTIONS TO CHAPTER OPENING MAZES — 119

INDEX — 121

PREFACE

This book contains a collection of mathematical puzzles, tricks, and oddities that have intrigued not only the author but also his friends and students over the years. Many have served as tension-breakers when used to interrupt a particularly involved lecture on some other topic in the author's classes during his college-teaching tenure. The math-oriented items included herein have proved of interest at the same time that they have afforded a degree of mental stimulus because of their curious aspects.

Many books have been written on various aspects of mathematical oddities, tricks, puzzles, and games. The items herein are not intended to be all-inclusive, because a volume of many more pages would be needed. Rather, the intent has been to emphasize those items in the literature (plus some original items by the author) that have proved of particular interest to the average nonmathematician as well as to the seasoned professional. Analytical comments, variations, and detailed explanations have been included wherever possible to provide a more thorough survey for those who like to delve a bit more deeply into these aspects.

I must acknowledge my production editor at Prentice-Hall, Barbara Palumbo, who through her own creativity and in keeping with the spirit of the book, developed the idea for and designed the mazes on the title page and chapter openings.

It is hoped that the reader will experience as much amusement from this collection as others have when first introduced to these interesting and often unusual math puzzles, tricks, and oddities.

MATTHEW MANDL

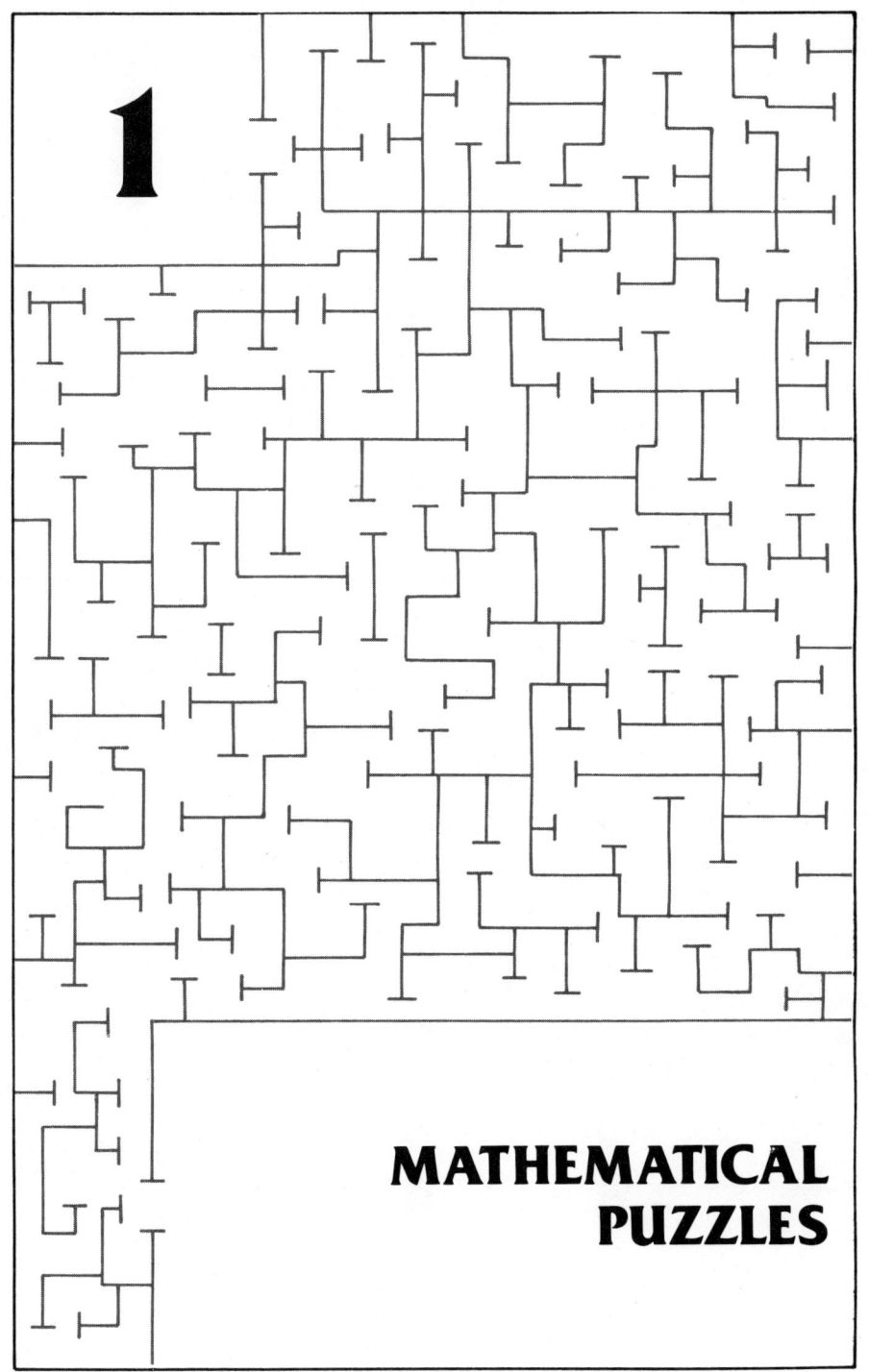

1

MATHEMATICAL PUZZLES

1-1. AN EXERCISE IN LOGIC

The puzzles in this chapter are of such a nature that they require a sense of logic as much as a general knowledge of arithmetic processes. For the most part the puzzles selected are those that deviate somewhat from the mathematical problem-solving type. As with all puzzles, there are two basic types, those that are solved with straightforward reasoning and those that require reverse logic. It is the latter type that presents the greatest challenge because the solution is usually obtained by an approach not usually thought of because it differs from that which commonly prevails.

A typical example of the reverse logic that must be applied on occasion is the classic nonmathematical puzzle of the multimillionaire who had only two children, each an owner of a prized race horse. When the rich parent died, an unusual will was found that stipulated all the fortune would go to the owner of the race horse that *lost* the race between the two offspring. Obviously, each time they held a race, both lagged behind as much as possible. Since neither horse managed to lose, they finally consulted a learned judge for advice. The judge solved their problem with one spoken bit of advice. Immediately each person jumped on a horse to race again, but now each urged the horse to run as fast as possible. What was the judge's advice? Although the judge's advice could almost be considered a trick answer to the puzzle, it is thoroughly logical in its perfect solution. The advice was "Trade horses and hold the race again."

Sometimes, however, the reverse logic is not as simple as the foregoing example but can become rather complex in the reverse-logic application.

Take, for instance, another famous puzzle wherein a person is traveling to New York and comes upon a crossroad section and notices that the directional sign for New York has fallen down. Two people approach and by their peculiar garb the traveler recognizes them as being from a nearby village with unique inhabitants. One part of the population always tells the truth and the other always lies. Each section dresses differently, one always in white and the other always in black. The traveler knows this but does not know whether the person dressed in white or black is the one that tells the truth. Since the two people are dressed differently, our traveler knows that if one of these persons is asked which path leads to New York, there would be no way of knowing if the answer is a lie or the truth. After some thought, the traveler asks one of the persons a simple question which, when answered, immediately discloses which path is the correct one. What is this question?

Here is an instance where the solution to the puzzle is again obtained by utilizing reverse logic. This is indeed what the traveler had to do to elicit an answer that was foolproof in determining the proper path to New York.

For convenience let us designate the townspeople as A and B. The traveler asks A the following question: "If I ask B if the left path goes to New York, would the answer be a *yes*?" If the left path actually is the one to New York and if A is the liar the traveler would hear a "no." If, however, A were the one telling the truth, there would also be a "no" to that question since B, being a liar, would say "no." Therefore, a "no" answer indicates the correct path has been pointed to by the traveler. Suppose, however, the traveler has pointed to the wrong road in asking A the question. Now if A is a liar and knew the path was incorrect, the answer would be "yes." If A were the one telling the truth, A would state that B would say "yes" because B would be the liar. Thus, if a "yes" answer is obtained it would indicate the traveler had pointed to the wrong path and thus now knows the correct path.

1-2. THE TWO COIN PUZZLE

This is a classic puzzle that has appeared numerous times in literature. For those who have not yet encountered it, it is still an excellent brain teaser. Some of the factors in logic discussed in Sec. 1-1 certainly apply here. The puzzle, simply stated, is this: I have two coins (U.S.) adding up to 55¢. One is not a nickel (5¢ piece). What are these two coins?

1-3. THE SIMPLE EQUATION

In this puzzle and throughout this text, the words *digit* and *numeral* each refer to a single unit such as 3, 7, etc. The use of the word *number*, however, indicates two or more digits making up a specific numerical value such as 28 or 697. This particular puzzle consists of the brief question: What equation using three identical digits results in 30 as an answer?

Sec. 1-9 What's Wrong Here? 5

1-4. THE BOUNTIFUL CHOICE

Which would you rather have, one million dollars or 1¢ the first day, 2¢ the next day, 4¢ the next day and continue doubling each amount until 30 days have elapsed?

1-5. TRICK QUESTION NO. 1

What number when multiplied by its reciprocal equals 1?

1-6. A FACTORIAL QUESTION

Would you rather have one million dollars or (using the factorial) 10! dollars?

1-7. RELATIVELY SPEAKING

In comparing the microseconds in a minute to the number of minutes in a century, which sum is the greatest?

1-8. DOUBLE TROUBLE

This is a two-question puzzle and is alphanumeric because numbers and letters must be found. Thus, in addition to the math solution, you must also find how many times the letter f appears. If you see only two or three fs, check again before looking up the answer at the end of the chapter. Here is the complete statement:

> When five digits are added, the sum is 7. Three of these five digits are 3, 2, and 1. Which two digits of many possible combinations can you find to satisfy the problem?

1-9. WHAT'S WRONG HERE?

A person visits a pawnbroker and pawns a $5.00 bill for $4.00 and receives the usual pawn ticket. He now sells the ticket to a friend for $3.00, thus receiving a total of $7.00 and a profit of $2.00. The friend redeems the ticket for the original $5.00 and hence makes a profit of $2.00. Who loses?

1-10. WHY ALWAYS 26262626?

When certain numbers composed of duplicate pairs of digits (such as 13131313) are multiplied by a specific number, the product is 26262626 (see also Sec. 4-10):

$$16161616 \times 1.625 = 26262626$$
$$13131313 \times 2 = 26262626$$
$$80808080 \times 3.25 = 26262626$$
$$6565656.5 \times 4 = 26262626$$
$$5252525.2 \times 5 = 26262626$$
$$4040404.0 \times 6.5 = 26262626$$

What are four additional examples using the same pattern of multiplicands and producing the same product based on the clues evident in the preceding examples?

1-11. THE SIMPLE CUBE ROOT

If the cube root of $1.371330631 = 1.111$, where must the decimal place be to yield a cube root of 11.11?

1-12. WHAT'S THE COST?

Three items were purchased, and for convenience we will designate them as A, B, and C. Item A cost four times more than B. Item C cost four times more than A. The total cost of all three was less than $2.00. All were bought and fully paid for with only two U.S. coins, neither of which was a penny. What were these two coins?

1-13. THE SQUARES HAVE IT

The difference between the square of two numbers is 12,200. When a "1" is added to either end of each number and each number is squared again, the difference is 1,222,000. What are the original numbers?

1-14. TRICK QUESTION NO. 2

Which is the correct statement? *The reciprocal of 8 is 0.120* or *The reciprocal of 8 equals 0.120?*

1-15. IT'S ALMOST IMPOSSIBLE

This puzzle is a new version of an old classic that requires full utilization of the reverse logic discussed in Sec. 1-1. A computer manufacturing firm needed a programmer who was a master logician. Finally, after many interviews, three individuals were left from the many who applied. In order to select the most talented, the three were given a special test simultaneously. Each was seated around a table, and each had a light that could be turned on with a push button. They were temporarily blindfolded, and a hat was placed on each person's head. Now the blindfolds were removed. All had the numeral 9 printed on the front of their hats. The numeral was not visible to the one wearing the hat, but the other two could see it.

The three were now told to press the button and light their individual lights if they saw a 9 on the hats of either or both of the other two. They were then also to try to apply logic to the situation and whichever individual was sure of the situation and could prove that the hat worn contained a 9, could stand up and announce this. At first it appeared impossible since each person's light was lit, and each person saw two 9s and apparently had no way of ascertaining whether or not his or her hat contained a 9. However, within a few minutes one person jumped up and pronounced positive knowledge of possessing a hat with a 9 on it. What was this person's logical explanation?

1-16. BRAIN TWISTER

Can you solve the following puzzle without using a calculator and only reading the question once? If you take one half of half a dozen and then add one half of the product that would result by multiplying the number by 6, would the final answer be 7, 12, 15, or 18?

1-17. WHAT PATTERN?

For the two columns of numbers below, there are many different numbers that can follow and be valid. Select one that will continue the pattern. The patterns differ for columns *A* and *B*.

A	B
10684	63012
13990	15015
22061	09921
13360	12890
96966	10101
43780	73011
49370	12027
12430	86506
50976	12345

1-18. UPSIDE-DOWN MESSAGES

On many occasions short articles have appeared in magazines illustrating how the calculator can be used to print out messages. Various numerals, when the calculator display is viewed upside down, appear as letters. The 7, for instance, appears as L, a 3 appears as E, 5 appears as S, and so on. As an example, perform the following operation on your calculator and read the answer by holding the calculator upside down.

$$\sqrt{14.953689} \text{ divided by 5}$$

We can tie the process into an anecdote. For instance, what did the person say after the race when he bet on a long shot? To find the answer, solve the following problem and read the results on a calculator held upside down:

$$(17530 + 5.5) \text{ divided by 5}$$

1-19. LETTERS IN NUMBERS

The spelling out of numerals and numbers such as 2, 3, 12, and so on gives us a variety of letters. We encounter a v in the numeral 5 and an x in the numeral 6 and so on. Hence, what is the first number encountered that has the letter d in it? Also, which is the first number that has both the letter d as well as the letter a? (Only single-word numbers count and numbers such as twenty-five and sixty-seven are not valid in this puzzle.)

1-20. IT SOUNDS COMPLICATED

We have avoided the purely mathematical problem-type puzzles herein because any math book contains numerous examples. The following, however, is about as simple a statement as possible for a math problem while at the same time making it a seemingly trick question. It is, however, a straightforward conventional statement of a typical math text puzzler: What number, when multiplied by the reciprocal of its square, equals one fourth the original number? (To simplify the problem additionally, here is a hint: The number is less than 10.)

ANSWERS TO CHAPTER 1 PUZZLES

1-2. The two coins are: a half dollar and nickel. Obviously, one of these is not a nickel—it is a half dollar.

1-3. The three digits and the manner in which they equal 30 are: $3^3 + 3 = 30$.

1-4. The choice should be 1¢ the first day, 2¢ the second day, etc. Doubling it each time for thirty days would then amount of $536,870,912.

1-5. Any number when multiplied by its reciprocal equals 1.

1-6. The factorial ten is the proper selection because it equals $3,628,800.

1-7. There are 60,000,000 microseconds in a minute but only 52,596,800 minutes in a century—including leap years.

1-8. For the first part of the question, the other two digits could be 0.5 + 0.5, or many others including 0.3 + 0.7. For the second part of the puzzle, there are five f letters in the question. Most people see only three because they overlook the f letters in the "of" words.

1-9. The person redeeming the ticket loses, since he must pay the pawnbroker for the item redeemed.

1-10. By doubling the multiplier of the last example (6.5), we obtain 13 × 2020202 = 26262626. Similarly, doubling the 13 we obtain 26 × 1010101 = 26262626. Also 52 × 505050.5 = 26262626 and 104 × 252525.25 = 26262626.

1-11. The decimal place must be moved over so it appears between the one and three: 1371.330631.

1-12. The ratio is $B = 1$, $A = 4$, and $C = 16$. The ratio of 1:4:16 would give us 21¢ if each unit represented 1¢. Doubling this value produces 42, doubling again is 84, and doubling again is 1.05—which is our answer ($1.05). It is invalid to double again because this would result in $2.10, which exceeds the stipulated $2.00.

1-13. The original numbers are 11 and 111.

1-14. Neither. The reciprocal of $8 = 0.125$.

1-15. To solve this puzzle, we must initially identify ourselves with one of the contestants. Let us designate the contestants as A, B, and C. Now assume you are contestant B who was the one that jumped up and claimed positive knowledge and proof that the hat worn had a 9 on it. Here is the logical reason: Initially, you, as B, see a 9 on the hat of A and C, and because each person is pushing the light button, it is obvious that each person sees a 9. You, however, do not know whether A is pushing the light button because he only sees the 9 on C's hat, or sees it on *both you and C*. Hence, you now reason in this logical sequence: Suppose A looked at you (B) and did not see a 9. Obviously, C would then also note that you had no 9 on your hat. However, since C is still pressing the light button it would be obvious that A's hat must have a 9 on it, and therefore A would immediately jump up announcing a conviction of proof. Since, however, A does not immediately jump up there must therefore be a 9 on your (B's) hat also. Thus, you, as B, have solved the logical answer to the puzzle.

1-16. The answer is 12 because one half of a half dozen is 3. When we add half of the product of 3×6, which is 9, we obtain 12.

1-17. For column A a 0 is always at the extreme right of the number only when a 3 appears somewhere in the number. For B, a 0 appears at the center of the number for every number whose sum of digits equals 12.

1-18. If the equations were executed properly, the first one should read *hello* and the second should read *I lose*.

1-19. The first *d* appears in the number *hundred*, while both a *d* and an *a* appear in *thousand*. (The latter is the first number word in which an *a* appears.)

1-20. The original numeral is 2.

2

PUZZLES IN THE GEOMETRIC

2

INTRODUCTION

The term *geometric* in the title of this chapter is used as a broad reference to a variety of puzzles formed by cutting, folding, or arranging paper or cardboard sections. Since the puzzles so created are mathematical in nature, they are included herein and serve as a brief departure from the tricks and puzzles involving numerals exclusively. Some of the puzzles in this chapter can be converted to tricks as, for instance, the ones discussed in Secs. 2-1 and 2-2. Show them to a friend and the illusions produced are that you have made a section of paper shrink or stretch mysteriously. Similarly, the Mobius ring discussed in Sec. 2-4 can be handed to a friend who is told that the ring can't be cut in half to form two rings. As proof, your friend is handed a scissors and attempts to cut the circle apart to form two circles.

2-1. REASSEMBLE AND GAIN

The puzzle shown in Fig. 2-1 has appeared in numerous versions in the past but is just as puzzling in one version as in another. The form can be laid out with a ruler on blank paper, but a graph sheet will simplify the construction. Thus, draw the figure in (a) and number the sections as shown. As illustrated, using squares of graph paper as the guide produces a section of 8 X 8 squares, or a total of 64 squares.

Next, cut the pattern apart along the lines shown in (a) and reassemble it as shown in (b). Now you will find that you have one side of 5 squares and the other side of 13 squares for a total of 65 squares! Where did the extra square come from? The secret lies in the fact that the diagonal sections formed from one corner to the other do not provide for a perfect fit with adjacent sections. Hence, along this diagonal length lies the extra space formed. The hypotenuse of the triangles are such that they do not meet perfectly when laid out in the manner shown.

14   Puzzles in the Geometric   Chap. 2

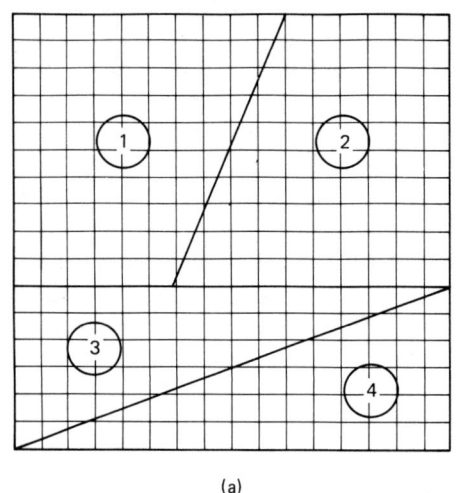
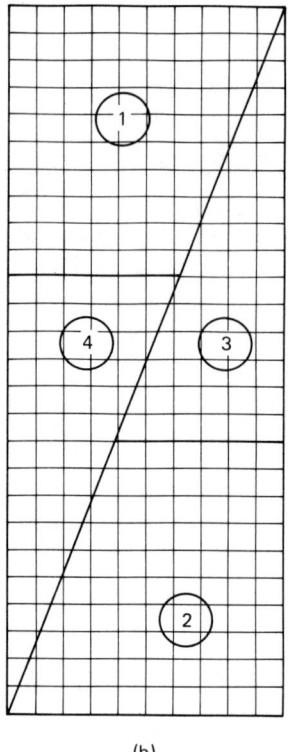

(a)   (b)

Figure 2-1

2-2. PAPER CAN'T BE STRETCHED

One of the most elementary and yet bewildering puzzles is simply formed by cutting two sections of paper into curved segments, as shown in Fig. 2-2(a). The illusion created by these two pieces of paper (or cardboard) is enhanced somewhat if different colors are used for the two sections. As shown in (a), if red and white pieces are used it can be proved that each is the same length when one is laid down facing the other. When, however, one is placed next to the other, as shown in Fig. 2-2(b), the upper piece will appear to have shrunk mysteriously while the lower one appears to have been stretched. The illusion becomes even more confounding if the white section is now laid alongside of the red with the white above the red. Now the white section seems to have shrunk while the red seems to have stretched.

Sec. 2-3　The Paper Cube Blowout

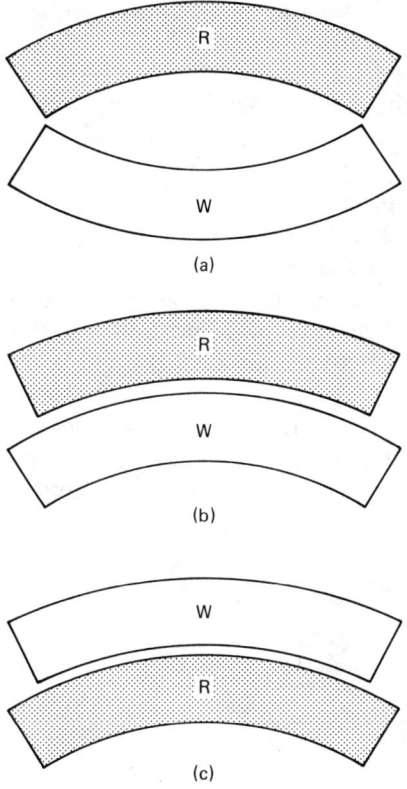

Figure 2-2

　　If each section is cut out using the same arc of curvature for each edge, the two pieces will fit adjacent to each other without an abnormal gap between them. Strangely, however, even a rough cut without using any drafting tools will still give the illusion of differences in length when one is placed next to the other. The length of the sections can vary without destroying the effectiveness of the puzzle. The segments can be made to fit in the palm of the hand, or they can be made as long as desired.

2-3. THE PAPER CUBE BLOWOUT

A rather novel geometric construction consists of folding a square sheet of paper in a certain manner so that air can be blown into the end result. This produces an inflated paper cube cleverly held together by its own folds. I was shown this item when I was a child and remember that after some practice I was able to construct it readily. Years later I recalled having made such a paper cube but didn't recall the folding procedure necessary to produce it. Subsequently, however, I ran across a science publication that contained mathematical paper folding and was surprised to find this old puzzle in it.

Once you have learned how to construct it, it becomes an interesting item to show to someone. In this respect, it is a convenient demonstration to do for a child on a rainy day. First you need to practice it a few times to memorize the peculiar folding processes. The explanation, of necessity, is more involved than the folding process itself. It is, however, an absorbing exercise in paper-folding techniques.

Initially obtain a perfectly square sheet of paper, preferably thin, and fold it several times along the lines shown in Fig. 2-3(a). Crease these folds inside and outside to facilitate the construction

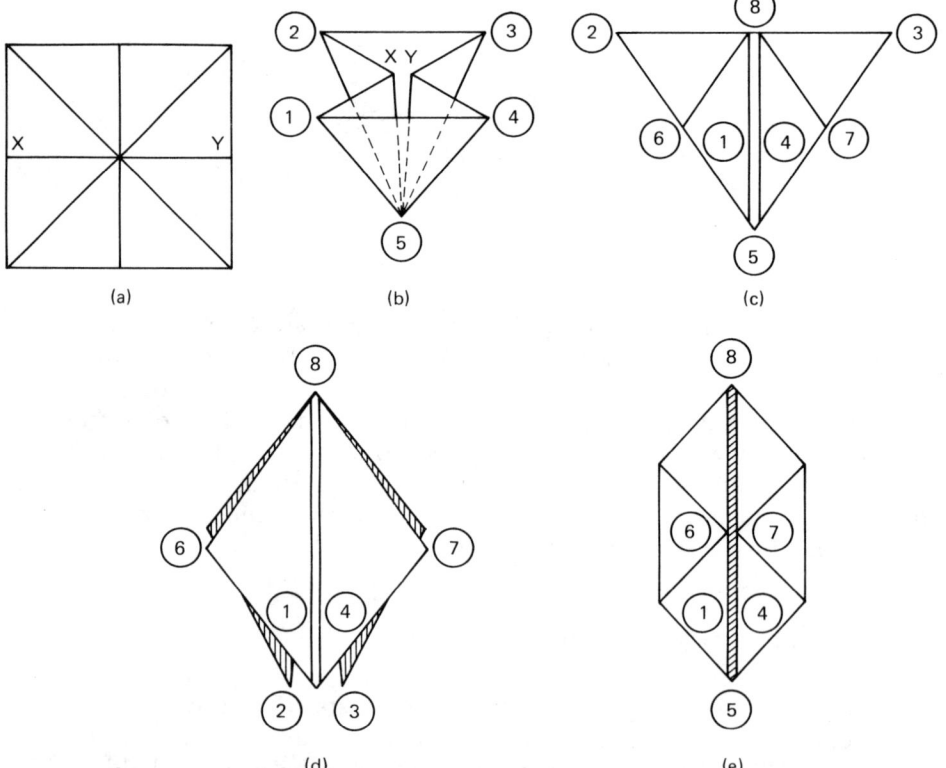

Figure 2-3

Sec. 2-3 The Paper Cube Blowout   17

that is to follow. Next, fold the paper in half once and hold the folded paper with the fold at the bottom. Now bring the corner folds marked *x* and *y* up toward the middle, as shown in Fig. 2-3(b). Bring points *1* and *2* together as well as *3* and *4*. Next fold points *1* and *4* down to meet point *5*, as shown in Fig. 2-3(c). Turn the paper over and fold points *2* and *3* down also to point *5*, as shown in Fig. 2-3(d). Now we have the top and bottom identified by *8* and *5* and the left and right sides identified by *6* and *7* as shown.

Note that the left and right side each consists of double sections. Fold the sections nearest you so the points meet at the center as shown in Fig. 2-3(e). Turn the paper over and again fold the *6* and *7* sides toward the center so the total pattern appears as shown in (e). Now refer to Fig. 2-4(a). Fold out the two bottom flaps as shown; turn the paper over and repeat for the other side. Now fold

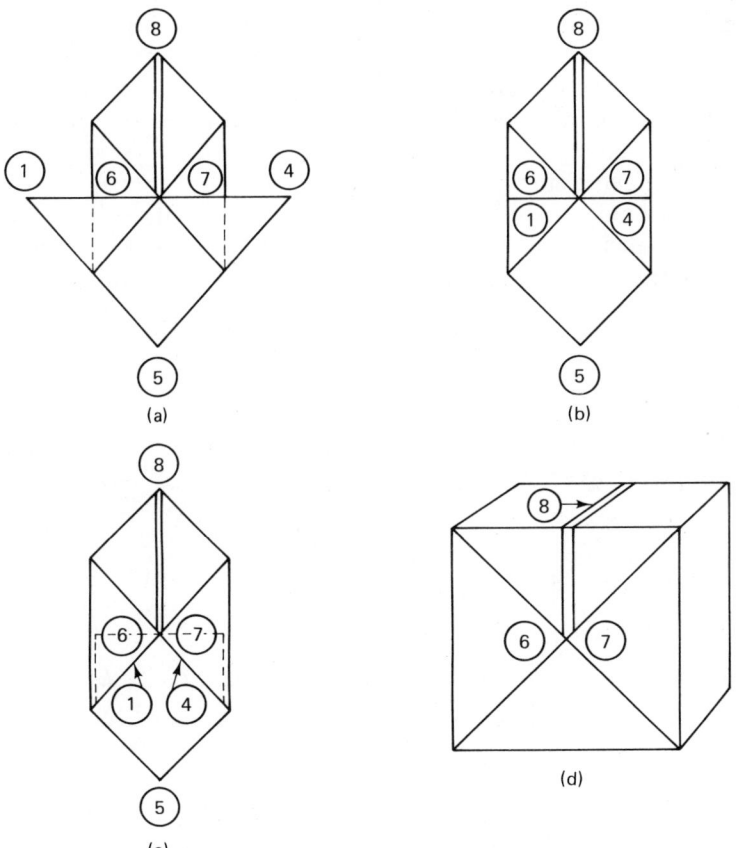

Figure 2-4

the flap ends that extend out in half to meet at the center as shown in (b). Turn over and repeat for the other side. Tuck the triangular wings inside the convenient slots that were formed beneath them to hold them in place. Turn the pattern over and repeat. Now the completed pattern should appear as shown in (c). Inspect the top and bottom, and you will note that one end has a small opening. Blow into this opening and the paper will expand into a perfect cube, as shown in Fig. 2-4(d).

2-4. THE MOBIUS RING

The famous Mobius ring has often appeared in mathematical and scientific publications. The device is named after August Mobius (1790–1868), the German mathematician who first devised it. The ring is easy to construct since it is composed simply of a strip of paper about a meter long that is given a single twist and glued or taped together at the ends, as shown in Fig. 2-5. Now we have a most peculiar device. It now has only one side instead of two, as is normally the case.

If we have a single length of paper, we state that it has two sides because we have to go over an edge to reach the other side. If we drew a pencil line on one side, we would have to cross the edge and continue the pencil line on the other. This is not the case with the

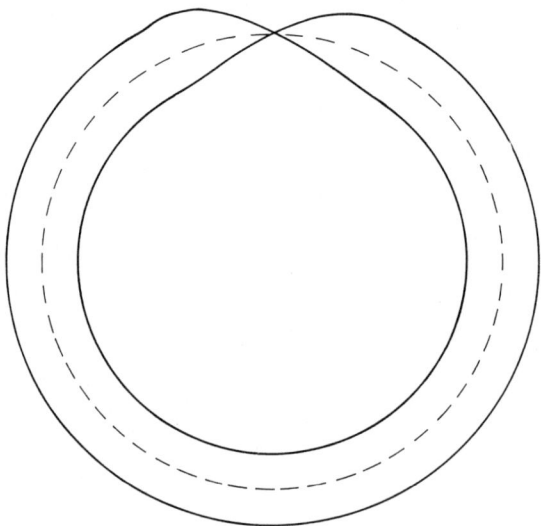

Figure 2-5

Sec. 2-5   Wheels within Wheels   19

Mobius ring, however, because if a pencil line were drawn along the center it would be found that if the line were continued it would reach the beginning of the line without having crossed over the edge. For additional proof, cut the paper ring in half along its circumference, as shown by the dotted line in Fig. 2-5. Normally we would obtain two rings by doing this, but the Mobius ring simply becomes another single ring, though of twice the original circumference.

2-5. WHEELS WITHIN WHEELS

A mathematical puzzle that is not too easy to solve can be formed by cutting out four rings of different sizes and marking off squares on the outer periphery of each ring, as shown in Fig. 2-6. The rings are then mounted on top of each other and joined at the center by pushing through a metal paper fastener of the type that has two prongs that can be spread out in the back. Next, the numerals shown are written on the spaces, after which the wheels are rotated. The puzzle is now ready, and the object is to move the wheels to the positions where the squares for each column of squares adds to 15.

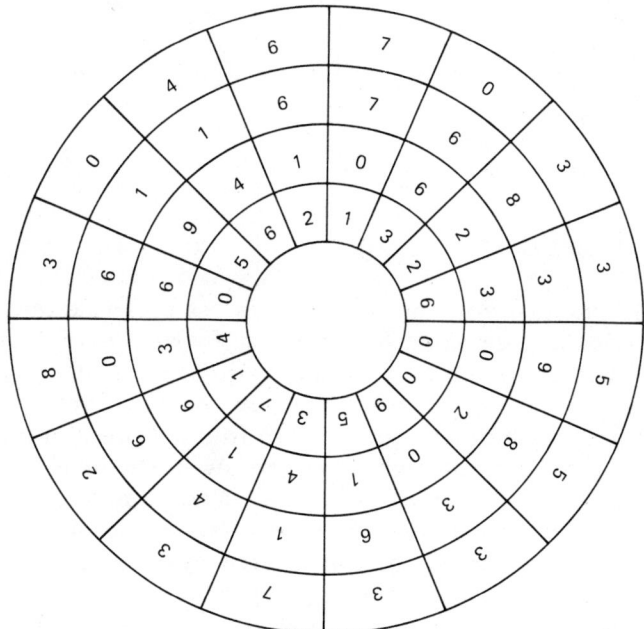

Figure 2-6

20   Puzzles in the Geometric   Chap. 2

The four-wheel puzzle shown in Fig. 2-6 is rather difficult to solve because of the numerous combinations possible. For children, a simple version can be made of three rings, as shown in Fig. 2-7, where each column of squares adds up to 12.

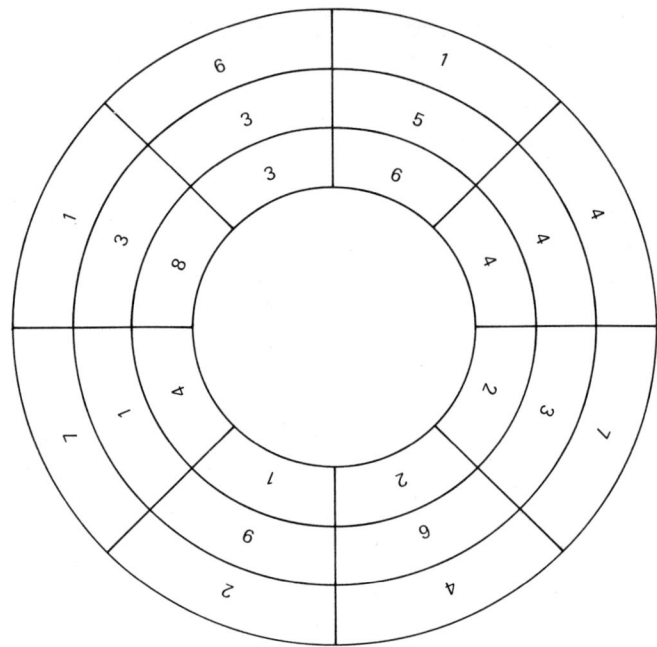

Figure 2-7

2-6. THE PRETZEL CAPER

With a little practice you can draw an almost perfect representation of two interlocking sections (see the last drawing of Fig. 2-8). The pattern, though simple in appearance, is quite difficult to imitate if you do not know the trick procedures involved. Following the illustration in Fig. 2-8(a), draw a square with four short extensions pointing up and down, left and right. Next, draw in the lines to form the figure shown in (b). The next two procedures complete the pattern, although they are a bit critical and may have to be practiced a few times. Draw in the curved sections shown in (c) to span a distance exactly equal to the width of the original square.

After the curved sections have been drawn in to appear as in (c), draw in the remaining curved sections illustrated by the broken lines

Sec. 2-6   The Pretzel Caper   21

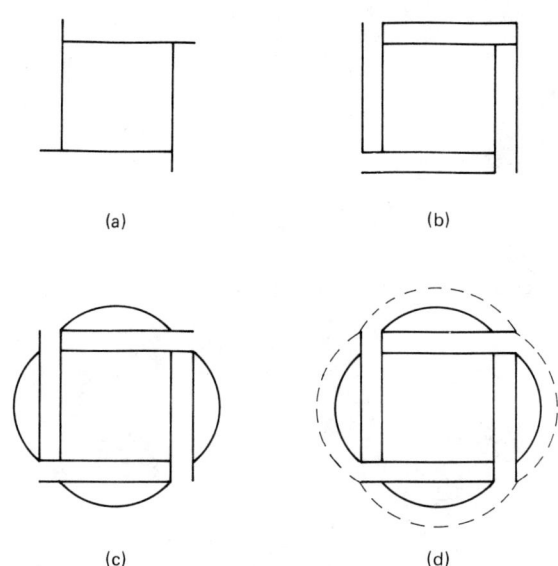

Figure 2-8

in (c). The lines are shown in dotted segments only to indicate the exact placement of the last curved sections. Draw these in with solid lines to form interlocking sections as illustrated. Now draw it for a friend and challenge him or her to duplicate it using only a pen or pencil.

3

MATHEMATICAL TRICKS

INTRODUCTION

Mathematical tricks have one aspect that differs considerably from the magic tricks used to create illusions. With the math tricks, the person to whom you are showing the trick is not the casual spectator, but must necessarily participate to bring the trick to a successful conclusion. Thus, the illusion is often more vivid or the basic aspects of the trick more mysterious. These factors apply for both the long or the short math tricks, although the added time of participation for the more lengthy ones heightens the suspense leading to the final effect. This is clearly exemplified in the first puzzle of this chapter, in which an apparently impossible procedure occurs because an answer is given to a five-tier addition problem before the participant discloses two more numbers to be used in the addition.

Some of the more elaborate tricks are somewhat tedious to fulfill if done without the aid of a calculator. Since the latter device is now so common, however, almost anyone can now participate in the completion of the tricks given herein without too much calculatory effort. Most of the tricks are fairly simple, although sometimes the explanations appear rather lengthy. However, these explanations are necessary to present a thorough groundwork in the trick, and complete familiarization is essential to assure successful presentation.

3-1. EARLY-SOLUTION TRICK

This is one of those tricks that seems impossible and hence always creates considerable interest when presented. You and the person to whom you are showing the trick will do a five-number addition, with the participant selecting three of the numbers, and you selecting only two. The very surprising factor here, however, is that you will write down the solution to the problem after your friend puts down the first number, but *before that person writes the second number!* The written solution can be folded and kept in possession of the participant while the trick is being performed. The layout is emphasized in Fig. 3-1 for convenience in visualizing the sequence. Note that there are three solid-line enclosures. The latter are the spaces to be filled in by the person involved in the puzzle. The dashed-line enclosures (two of them) are the only places where you fill in numbers.

Mathematical Tricks   Chap. 3

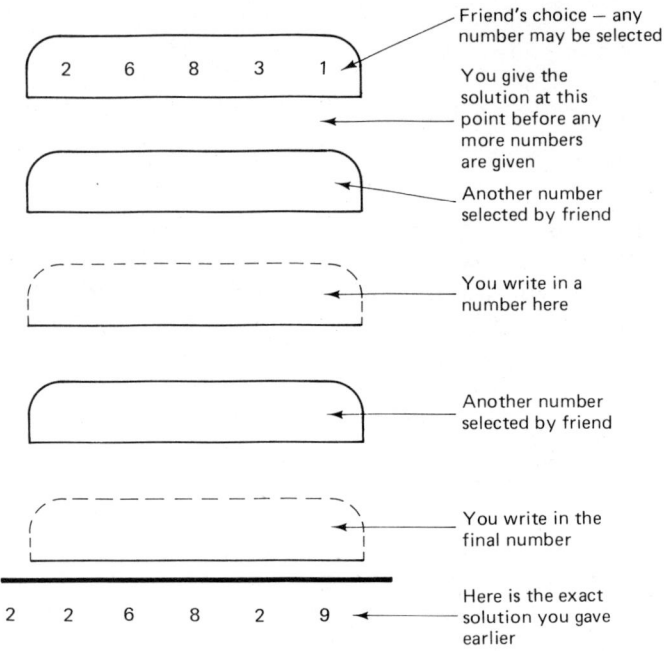

Figure 3-1

Initially, the person being shown the trick writes a number in the top section as shown. Assume the number 26831 has been selected (see Fig. 3-1). Now all the remaining numbers must have the same number of digits. (This is the only strict rule to be observed.) You immediately write down the solution (226829) and keep it aside until the conclusion of the trick. Now your friend selects another number, perhaps 15764. Next it is your turn, and you write in a series of numbers. Here is the problem up to this point:

$$
\begin{array}{ccccc}
2 & 6 & 8 & 3 & 1 \\
1 & 5 & 7 & 6 & 4 \\
8 & 4 & 2 & 3 & 5 \\
\end{array}
$$

The foregoing already contains the clue to the trick. Each numeral that you write in must add up to the 9 with the number right above it. Thus, at the extreme right, your friend had written a 4 and you placed a 5 beneath it, for a sum of 9. Similarly, in second place there is a 6 and you write a 3 below it, again for a total of 9, and so on across the number. Now, suppose your friend writes 50928 for the third number. Again you write your number below this so that each numeral when added to the upper again equals 9. Now we have

Sec. 3-1   Early-Solution Trick   27

completed the problem and the answer coincides with the one you gave after the first number was written by the participant and before the latter added the second:

$$
\begin{array}{r}
2\ 6\ 8\ 3\ 1 \\
1\ 5\ 7\ 6\ 4 \\
8\ 4\ 2\ 3\ 5 \\
5\ 0\ 9\ 2\ 8 \\
4\ 9\ 0\ 7\ 1 \\
\hline
2\ 2\ 6\ 8\ 2\ 9
\end{array}
$$

The most unusual aspect of this trick is that the same answer would have been obtained regardless of the second and third number choice of the participant, as long as you made sure the selections plus yours below them added up to 9 for each pair as illustrated. The only way a new solution is formed is to change the first number written down. Before we detail the method for obtaining the answer, let's assume your friend had selected the following numbers for the second and fourth columns. Your numbers are included to show the new ones necessary to obtain the 9 value sum. Note the solution is the same as the initial one shown earlier.

$$
\begin{array}{r}
2\ 6\ 8\ 3\ 1 \\
5\ 6\ 5\ 1\ 9 \\
4\ 3\ 4\ 8\ 0 \\
1\ 2\ 3\ 2\ 1 \\
8\ 7\ 6\ 7\ 8 \\
\hline
2\ 2\ 6\ 8\ 2\ 9
\end{array}
$$

The secret in finding the solution after the first number is given is to subtract 2 from the number, and add the 2 to the left end of the number. Thus, since 26831 was selected we subtract 2:

$$
\begin{array}{r}
2\ 6\ 8\ 3\ 1 \\
-\ \ \ \ \ \ \ \ 2 \\
\hline
2\ 2\ 6\ 8\ 2\ 9
\end{array}
$$

If the number first selected was a six-digit one and consisted of 407814, the solution to the whole problem would be:

$$
\begin{array}{r}
4\ 0\ 7\ 8\ 1\ 4 \\
-\ \ \ \ \ \ \ \ \ \ 2 \\
\hline
2\ 4\ 0\ 7\ 8\ 1\ 2
\end{array}
$$

The matter becomes a little tricky if the number selected is one ending with one or more zeros such as 18100. Now the solution would be:

```
  1 8 1 0 0
—         2
_____
2 1 8 0 9 8
```

Thus, if that initial number had been selected, the solution would have been sufficiently different from the initial number so that it would have been difficult for the participant to fathom what method you used to find it. Again, regardless of the numbers selected for the second and fourth rows, the answer would remain the same when you convert the numeral pairs to equal 9 across the entire row as previously detailed.

As a final example, assume the first number given is 520629. You give the solution as 2520627. Next 415268 is given, and your choice is then 584731. Finally, 360051 is given and your choice is 639948. Now we have the following, and the answer coincides with that given initially.

```
  5 2 0 6 2 9
  4 1 5 2 6 8
  5 8 4 7 3 1
  3 6 0 0 5 1
  6 3 9 9 4 8
_____
2 5 2 0 6 2 7
```

3-2. SECRET NUMBER REVEALED

Some mathematical tricks are performed more conveniently using a calculator to expedite the arithmetic processes. The next several tricks fall into this category, although if someone is adept in addition and multiplication, perhaps the end result will be more surprising. The following trick has two versions and both are unusual because not only is a secret number selected, but a numeral selected at random is subtracted from it before the remaining steps of the mathematics are completed.

Ask someone to select any two-digit number (without telling you what it is) and immediately subtract from the number any single

Sec. 3-2 Secret Number Revealed

digit (from 1 to 9). Now ask the person to multiply the remainder by 9 and add the original two-digit number (before the single digit was subtracted from it). The person now shows you the final sum without disclosing the original numbers. You mentally add the last number of the end result to the first two numerals to find the original two-digit number. As an example, suppose the number 47 had been selected, and the number 4 was subtracted from it. Now the remaining 43 is multiplied by 9 and the product is 387. To this product we add the original number of 47:

$$\begin{array}{r} 387 \\ +47 \\ \hline 434 \end{array}$$

This is where you add the last numeral (4) to the center numeral (3) and now have the original number of 47. Curiously, the same answer is obtained regardless of the single digit subtracted from the original number. Had the digit been 6 instead of 4 the process produces a new sum but the final answer is still 47.

$$\begin{array}{l} 47 - 6 = 41 \\ 41 \times 9 = 369 \\ 369 + 47 = 416 \\ 41 + 6 = 47 \end{array}$$

The trick can also be performed in a three-digit version of the original number. Again, however, the participant subtracts a single digit from it before multiplying by 9 and adding the original number. Now, however, the final sum will have four digits, and you must add the last two digits to form the true first place digit of the three-digit number. The following example will illustrate this factor.

Number selected: 456
Digit selected: − 3

$$453 \times 9 + 456 = 4533$$

The 4533 is thus modified by adding the last two digits, and the resultant 6 becomes the first place digit of the original sum: 45 with added-on (3+3) = 456.

Similarly, had 789 been selected and 6 subtracted from it we would have: $783 \times 9 + 789 = 7836$. Adding the last two digits gives us 9, and putting this in first place gives us the original number of 789.

3-3. FAST ANSWER TRICK

This trick is related to the Early-Solution Trick discussed in Sec. 3-1, although much shorter in presentation. Have someone write any number secretly. On a separate piece of paper ask the participant to again write down the same selected number, but now subtract 4 from it and also place a 4 to the left end of the remainder. Thus, if 56 were selected, subtracting the 4 gives us 52 and placing the 4 in place at the left gives us 452. This new number is put aside and the participant again picks up the first paper with the original selection. You, however, would not know how many digits were in the number selected. Hence you next ask how many digits are in the first number selected, without of course disclosing the original number. If the selected number had two digits, ask the participant to add 396 to it and the resultant sum will coincide with the amount on the paper put aside earlier.

If the original number had three digits, you ask that 3996 be added. Assume, for instance, that 576 had originally been selected. Subtracting a 4 produces 572, and placing the 4 at the left gives us 4572, which is put aside (and which will coincide with the final answer). Now the original number of 576 has 3996 added to it:

$$\begin{array}{r} 576 \\ +\ 3996 \\ \hline 4572 \end{array}$$ (same as modified original)

If the original number had 4 digits, the number to be added at the end is 39996, if 5 digits, add 399996, etc. In all instances, however, the 4 is subtracted and added to the left end to form the final answer before proceeding. Note that the number added at the end of the trick always consists of a 3 at the left, several 9s in the center, and a 6 at the right in first place. Total digits are always 1 more than the number of digits in the original number. Thus, if the original number selected were 23671, the subtraction and placement of 4 initially produces the following:

$$\begin{array}{r} 23671 \\ -\ \ \ \ \ 4 \\ \hline (4)23667 \end{array}$$

Because five digits were involved, the number to be added must contain 6 and hence is 399996, which again gives us the answer found earlier.

                23671
              + 399996
              ───────
                423667

3-4. THE MYSTERY ANSWER

Tell someone to use his or her calculator and secretly punch in any number with three similar digits (such as 555 or 777). Now have the person multiply the selected three-digit number by any number of any magnitude without telling you the multiplier selected. Next you request the product be multiplied by 6 and again multiply the new product by any other number selected at random. Now the digits of the answer appearing on the calculator are to be added together, such as 18954 would produce a sum of 27. If a two-digit number is obtained, these are again to be added together. Now you announce the final answer, which is 9.

This trick is related to the oddities of nines discussed in Chapter 4 and is also related to the process discussed in Sec. 4-4. The peculiar characteristic of this trick is that the participant selects the initial number, also chooses the multipliers, and only uses one multiplier suggested by you—the 6. The only limit to the process is that imposed by the calculator, since the total value soon increases considerably if a large-value multiplier is selected. As an example, note the magnitude achieved in the following:

    Number selected: 777
    Multipliers selected: 54 and 579
    Multiplier given by you: 6
    Hence: 777 × 54 × 579 × 6 + 145762092

Thus, in the example just given, the product already exceeds the usual eight-digit capacity found in most lower-priced calculators. Note that the sum of the digits, however, still gives the proper answer. The sum of 145762092 = 36 and these in turn add up to 9.

3-5. ANOTHER QUICK ANSWER TRICK

In contrast to the trick detailed in Sec. 3-4 where the multiplicand had to consist of similar digits, an interesting version that has no such requirement is the one discussed in this section. Again, a calculator would be handy, although if a low-value multiplier is selected only a few mental calculations may be needed. For this trick ask someone to select any number (no restrictions on types of digits, number of digits, or magnitude). Now the selected number is to be multiplied by any one of the following multipliers the participant chooses to select: 6.75, 13.5, 22.5, or 45. Subtract 5 from the product obtained and throw out any 0s, 9s, and decimal points. Add the remaining digits and the final sum will always be a 4.

As an example, assume the product 60117 was selected with the 45 as a multiplier. The product is:

$$(60117 \times 45) = 2705265$$
$$(\text{less } 5) \quad - \quad 5$$
$$(\text{product}) \quad 2705260 \quad \text{(the sum of which is } 22 = 4)$$

Note the following example in which a decimal point appears after the initial multiplication:

```
          625458   (selected multiplicand)
    ×       6.75   (selected multiplier)
       ─────────
        4221841.5  (resultant product)
    −         5.0  (subtract 5)
       ─────────
        4227836.5  (throwing out decimal point leaves
                   42218365, the sum of which is
                   31 = 4)
```

3-6. THE HIDDEN PRODUCT

This trick is related to the puzzle described in Sec. 4-2 in the next chapter. The trick involves several multiplication processes, and the product obtained is sufficiently far removed from the original number thought of to prevent the participant from fathoming how you found the original number used for the multiplicand. To perform this trick ask someone to think of any three-digit number. (Preferably the number should be entered into a calculator for convenience, since several steps of multiplication are used.) Now ask the person to multiply the number by 55, then by 7, and finally by 0.65. Ask to see the resultant product and without disclosing what you do, multiply it by 4. Now you will obtain duplicate numbers, and can select either one as the original number used. Two examples follow:

Number selected: 126
126 × 55 × 7 × 0.65 = 31531.5
Multiplying the latter by 4 gives us 126126, of which either the first three or last three digits is the original number.

Number selected: 921
921 × 55 × 7 × 0.65 = 230480.25
Multiplying the final product by 4 gives you 921921, hence the original number was 921.

3-7. THE NUMERICAL SHELL GAME

To perform this trick you need four identical squares of cardboard. These can be any size, although for convenience they can be approximately the size of a book of matches. On one of the cardboard sections write the numeral 1, on the next card the numeral 2, on the following card the numeral 3, and on the last card the numeral 9. Cardboard is preferable so these numerals do not show through to the back. These cards are then shuffled and laid face down on the table. In demonstrating this trick to someone you then mention that the numeral 9 has a curious affinity for humans and will automatically end up with the person to whom the trick is shown. You then ask the person to point to two of the four cards and finally to one of the two left. Magically the remaining card is the one with the 9 on it. Thus, this trick differs from those in which the magician finds a specific card. Here the participant appears to have unconsciously selected the one with the 9 on it.

The manner in which this trick is performed is by a standard magic procedure known as forcing. Initially when you lay the cards on the table make sure *you* know the exact location of the card with the 9 on it, but at the same time do not disclose its location to the person to whom you are showing the trick. If the two cards that were pointed to should be the 1 and 3, you remove these cards, then ask that another card be pointed out. If this new selection is the card with the 2 on it, remove it also. Now only the card with the 9 on it is left, and you then turn this up and state that the affinity of the 9 has proved itself. If, however, the person initially pointed to the cards with the 2 and 9 on them, you *leave them on the table* and withdraw the 1 and the 3 card. The participant, not having seen the trick before, has no reason to believe that this is not the normal procedure. If the participant points to the 9 card of the two that are left, you remove the 2 and leave him with the last card he pointed to as proof of the magic of the 9 card.

34   Mathematical Tricks   Chap. 3

Because you may withdraw the first two cards that are pointed to at one time or perhaps find it necessary to leave them on the table, it is advisable to do this trick only once. If a trick is done a second or third time the participant usually notes the disparity in the selection procedure and the illusion is lost.

3-8. HOW IS IT DONE?

In this trick we use the square grouping shown in Fig. 3-2. Ask someone to think of any number between 1 and 15, then indicate to you which blocks contain the number, such as *A, C, D*, etc. You immediately announce the number that was thought of. The trick is that you mentally add up the numbers shown in the lower right square of each block designated. If, for instance, the number 7 was thought of, you would be told that the number appears in blocks *A, B, D*, and *F*. The lower right square numerals are 1, 0, 2, and 4, adding up to 7. Similarly, any other numbers selected will be found by adding the digits found in the lower right square.

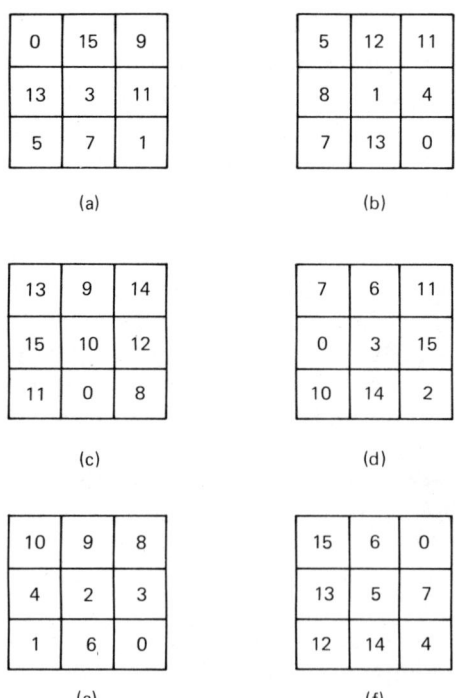

Figure 3-2

Sec. 3-11   Trick Game Is in Your Favor   35

3-9. CALCULATORS DON'T LIE

Ask someone to punch a number on a calculator, multiply it by its square, then multiply the product by the original number. Now you are handed the calculator, which displays the final product. You punch a key twice and immediately disclose the original number selected.

The key punched twice to find the original number is the square-root key. As an example, suppose the original number selected was 8. Multiplying this by its square is $8 \times 64 = 512$. Multiplying this by the original number (8) yields 4096. To find the answer you push the square-root key once and obtain 64, and the square-root key again gives you 8.

3-10. THIS ONE'S EASY

Tell someone to think of a number, multiply it by 3, then multiply it by 2; divide it by 12 and add 5. Now you look at the answer and immediately announce what the original number was. The original number is found by multiplying the answer that was shown to you by 2 and subtracting 10. As an example assume the chosen number was 359. When this is multiplied by 3 and 2 we obtain 2154. When the latter number is divided by 12 and 5 added we obtain 184.5. This is the number shown to you. Now you secretly multiply this by 2 and subtract 10 to yield 359.

3-11. TRICK GAME IS IN YOUR FAVOR

Here is a little game you can play that you will win most of the time, and on occasion all of the time. In this game we use 13 objects, which can be pennies, buttons, paper clips, and so on. Only two players are involved: you and your opponent. The rules are simple: each player takes a turn removing 1, 2, or 3 pieces and whoever is left with the last piece loses. Neither player may skip a turn. Players take turns going first.

If your opponent goes first you can't lose. The trick is to make sure the number of pieces picked up by your opponent and you at each turn always equals 4. Thus if your opponent picks up 1 piece, you pick up 3. If your opponent picks up 2 pieces, you pick up 2, and if 3 are picked up, you pick up 1. As an example, if your opponent picks up 2 initially, you pick up 2 leaving 9. If your opponent picks up 1 the next time, then you pick up 3 leaving 5. Now it is obvious that you can't lose because if only 1 is picked up, you will pick up 3 leaving the last 1. If your opponent picks up 2, you will pick up 2, and so on.

If you go first and your opponent does not know the secret you can win most of the time. Thus, in your initial turn only pick up 1 piece. If your opponent now picks up *any less than 3* you can again win. But if by chance 3 are picked up, you can still win because in your next turn you will again pick up 1. If your opponent now picks up 2 the game is lost, since you now control the selections. Thus on your next turn pick up 1 again. This now leaves 5 pieces and consequently your opponent can't win because if 1 is taken, you will take 3 and 1 is left for your opponent. If your opponent takes 2, you will take 2 leaving only 1, whereas if 3 are taken, you will take 1 leaving only 1.

If your initial selection is 1 and your opponent selects 1, you will select 2 for your next turn and immediately you will have control of the game again because a total of 4 pieces have been picked up. If your opponent takes 3, your next selection will be 1, and if 2 are taken your next selection will be 2, leaving the last 1 again. The only way you would lose when you start first is if your opponent picks the right number of pieces at each turn to add up to 4 as explained at the beginning of this discussion (wherein you go first). The chances of your opponent doing this, however, are not too good unless the secret of the game is known.

3-12. FOR YOU IT'S EASY

This is a trick that is much more work for the other person than it is for you. In this trick you ask someone to think of a number (or preferably punch it on a calculator). Next the volunteer is asked to multiply the selected number by 3, and again multiply it by 3. Now ask the participant to add up the sum of the digits in the answer and double the amount so obtained. Finally, 20 is added to the sum. Now you tell the person that the final answer is 38. (It always is.) As an example, assume the participant selects 238. When this is multiplied by 3 and again by 3, the product is 2142. Adding these numerals together gives us 9 and doubling this produces 18. Adding the final 20 gives the new sum of 38.

3-13. ROUND-ABOUT TURN-AROUND

As with the trick in Sec. 3-12, this one also always produces the same sum, although in this case it is 1089. However, in this trick there are a couple of restrictions. The participant must take a number above

300 but less than 988, and it should have three dissimilar numerals. The person reverses the sequence of numerals and subtracts the smaller from the larger. This new number is also reversed in sequence and added to produce the final sum of 1089. As an example, assume the participant selects 752. When this number is reversed and subtracted the number 495 is obtained, and this is again reversed and added as follows:

$$\begin{array}{r} 752 \\ -257 \\ \hline 495 \\ +594 \\ \hline 1089 \end{array}$$

This same trick can be performed with four numerals, in which case the final answer would always be 10890 as shown in the next example, in which the initial number is 7512. The limits of the initial number should be set between 1234 and 9876 while also using numbers with no duplicate numerals in them.

$$\begin{array}{r} 7512 \\ -2157 \\ \hline 5355 \\ +5535 \\ \hline 10890 \end{array}$$

3-14. TWO TRICKY TRICKS

In these two tricks a calculator is essential to reduce tedious math. Each trick has a similar pattern and involves a lengthy mathematical process that ends by your disclosure of an original number selected by the participant. In the first trick the participant selects any number and punches it into the calculator. Now you ask that 222 be added to the original number. Next the new sum is multiplied by 3, and the original number is subtracted from the product. Now ask the participant to subtract the number 666 from the answer and then to hand the calculator to you. You immediately divide the displayed number by 2 and hand the calculator back with the original number now showing. As an example, assume the original number selected was 153. The mathematical process then becomes:

$$
\begin{array}{r}
153 \\
+\ 222 \\
\hline
375 \\
\times\ \ \ 3 \\
\hline
1125 \\
-\ 153 \\
\hline
972 \\
-\ 666 \\
\hline
306
\end{array}
$$

(divide by 2 = 153)

A variation of the foregoing is performed without using any three similar numerals such as the 222 and 666 in the preceding. In this version 232 is added initially and 696 subtracted finally as in the following example:

$$
\begin{array}{r}
297 \\
+\ 232 \\
\hline
529 \\
\times\ \ \ 3 \\
\hline
1587 \\
-\ 297 \\
\hline
1290 \\
-\ 696 \\
\hline
594
\end{array}
$$

(divide by 2 = 297)

For the following trick a volunteer adds any number to 999. Next 351 is subtracted from the sum and then 537 is subtracted from the answer. When the final number is disclosed to you, mentally subtract 111 and you have the original number. As an example, assume the selected number was 247. The mathematical process is then as follows:

Sec. 3-14   Two Tricky Tricks

$$
\begin{array}{r}
999 \\
+247 \\
\hline
1246 \\
-351 \\
\hline
895 \\
-537 \\
\hline
358
\end{array}
$$

(you subtract 111 and get 247)

4

ODDITIES IN MATHEMATICS

INTRODUCTION

There are many mathematical processes that yield rather curious results. Most of them consist of problems or equations that produce unexpected answers. Some aspects of such mathematical oddities have already been explored in earlier chapters where they are utilized as puzzles or as math tricks. The examples in this chapter could also be used as puzzles or as tricks, although they stand on their own in terms of peculiar results obtained for the math sequences illustrated.

4-1. RECURRING SEQUENCE OF ONES

There are several math processes that yield a series of 1s for the answer, notably the square root of a sequence such as 123454321. The latter is termed a *palindrome* and can be expanded up to 9 (12345678987654321). The peak of the numbers gives a clue regarding the answer. Thus, the square root of 1234321 = 1111, and the square root of 12345654321 yields 111111 because the peak is reached at 6. With the earlier example where the sequence ascends to a 9 before it descends, the answer is nine 1s. Obviously, the square of any number such as 11111 or 11111111 will produce a corresponding palindrome.

Another process that yields a series of 1s is to take any number of 8s and divide successively by 1.6 and 5. Thus, 8888 divided by 1.6 and 5 produces 1111, whereas 88888888 produces 11111111. A more obvious procedure of this type is to divide by 4 and 2. Because the division by four yields a succession of 2s, an additional division by 2 produces the 1s (see also Sec. 4-4).

A similar result can be obtained by grouping any desired number of 9s and dividing successively by 3, 5, and 6. Thus, 99999 divided by 3, 5, and 6 produces 1111.1. It is easy to set up random-number examples by simply multiplying a series of 1s by the required amount. Raising 1111 to the third power gives us 1371330631. Hence, the cube root of the latter will produce a series of 1s. Such a contrived example, of course, does not rival the oddities mentioned earlier. The examples using 8s and 9s yield 1s regardless of the number of 8s or 9s employed.

A process yielding 1s for the answer that has some interesting aspects is the division of 111111 by 7, 13, and 11. The result is a three-digit group of 1s. When, however, we insert a 2 into the group of 1s (1112111) and again divide by the previous sequence, we obtain four 1s. Similarly, placings two 2s into the middle of the six 1s (11122111) yields an end result of 11111. Every time an additional 2 is placed in the middle of the original six 1s, an additional 1 appears in the quotient. We can, of course, do this problem in reverse. Thus, if we multiply seven 1s by 7, 11, and 13, we get a product of 1112222111.

4-2. THE ORIGINAL NUMBER REAPPEARS

If you multiply any number by 3, then again by 3, and then add the original number to the product, you will get an answer containing the original numerals, although of a different value unless multiplied by 0.1. Thus, 457 × 3 × 3 + 457 = 4570. Had the number been 23.69, the answer would have been 236.9. Multiplying by 0.1 would produce the original value. Thus, (23.69 × 3 × 3 + 23.69) × 0.1 = 23.69.

Another curious process regarding the reappearance of the original number involves the 7, 11, 13 sequence of divisors used in Sec. 4-1 to obtain the series of 1s in the answer. For the reappearance procedure the original number is repeated in combination with itself to form a double number.

For instance if the number 367 is selected, we follow it immediately with a repeat of the original number (367367). Now, if we divide this combination by 13, 11, and 7, we obtain the original number of 367. This procedure produces correct results only if three numerals are used before duplication. Thus for 286, the duplication is 286286 and is valid, whereas only two numerals, such as 74, duplicated to 7474 will not produce the original number.

4-3. THE STRANGE ODDITY OF NINES

There are a number of peculiar results obtained when 9s are involved in calculations. Some of the strange characteristics associated with 9s have already been touched on before, notably in Sec. 3-1 in the Early-Solution trick. One of the most fundamental oddities is that any number when multiplied by 9 will have a product consisting of

Sec. 4-3 The Strange Oddity of Nines

numerals that add up to 9. As an example, if 32 is multipled by 9 we obtain 288 and adding the numerals gives us 18, which yields 9 when added together again. Similarly, multiplying 2318 by 9 gives us 20862, and adding these numerals together produces 18, which then adds up to 9 again. Even a negative number does not disturb the pattern; multiplying 2358 by −9 gives a product of −2115, which adds up to 9.

Take any number with dissimilar numerals (above 11), reverse the number and subtract. The remainder will have numerals that add up to 9. If, for instance, we select 72 and reverse this, the subtraction is 72 − 27 = 45, the numerals of the latter giving us 9. Similarly, 9742 reversed gives us 2479 and the subtraction yields 7263, which adds up to 18 and hence finally to 9. Numbers such as 202 are invalid, since the reversal gives us the same number. Similarly, 818 or 444 are invalid.

Another example of the oddity of nines is found for any number above 9. When its digits are added and then subtracted from the original, it produces an answer having digits that add up to 9. Thus, 651 when added gives us 12, and 651 − 12 = 639 = 18 = 9. Similarly, 1568 produces 20 when the digits are added, and subtracting this from the original yields 1548 = 18 = 9. An interesting example is 12345 − 15 = 12330 = 9. Actually, you can subtract the larger number from the smaller and get a negative answer, but the addition of the numerals in the answer still gives us 9:

$$2406769 - 9676042 = (-7269273) = 36 = 9$$

If the multiplicand consists of numerals that add up to 9, the product will always consist of numerals that add up to 9, regardless of the multiplier: 63 × 5 = 315 = 9. Also, 54 × 2 = 108, and 72 × 8 = 576 = 18 = 9, or 81 × 3 = 243, and so on. When the sum of the numerals of a number such as 782 is used as the multiplier, the product will again have numerals that add up to 9. Thus, 732 × 12 = 8784 = 27 = 9. Similarly, 2364 × 15 = 35460 = 18 = 9 again!

Multiplication of any number by 6 then by 3 yields a product that gives us 9 when the numerals are added, as 23 × 6 = 38 × 3 = 414 = 9. Similarly, multiplying by two or three successive threes also produces the end result, as 23 × 3 = 69 × 3 = 207 = 9. Multiplying by 2 and then by 7 also produces this result, as does 4 and 5. One would tend to believe that the reason for this is that the successive multipliers (6 and 3) add up to 9, as does the 3 × 3 × 3 sequence. This reasoning is invalid, however, because multiplying by 4 then 5 (the sum of which is 9) does not produce numerals adding up to 9. As an example, note that 12 × 4 = 48 × 5 = 240. Multiplication by 7 and 2: 4 × 7 = 28 × 2 = 56 does not produce the required results either, as for instance (4 × 7) × 2 = 56.

Of interest also are the following examples of recurring 9s:

$$\sqrt{999{,}999} = 999.9995$$

$$\sqrt{99{,}999{,}999} = 9999.99995$$

4-4. SAME ANSWER REPEATED

In Sec. 3-4, the mystery answer trick produced a sum-of-digits answer of 9 regardless of the several multipliers (as well as the multiplicand) that were selected by the participant. This was valid as long as one multiplier selected by you was used (the 6). This mathematical situation is rather unique, because the same results could be obtained by using a 3 as one of the multipliers. Thus, we can set down a rule regarding this as follows:

> Take any number made up of three similar digits (222, 777, etc.). Multiply the selected three-digit number by 3. Then multiply by any number of additional successive multipliers of any magnitude. The final sum of the digits in the product will always be 9.

Actually, the multiplication by 3 can be employed at the beginning, during the middle, or at the end of the succession of multipliers without disturbing the end result. Here are a few examples:

444 × 7 × 3 × 21 × 276 = 54041904 = 27 = 9

222 × 28 × 4 × 3 × 92 × 31 = 212736384 = 36 = 9

This procedure is also valid for any six-digit number composed of alike numerals:

777777 × 3 × 2 × 56 = 261333072 = 27 = 9

The procedure will also work for a nine-digit number, a twelve-digit number, etc., as long as all digits are the same. With the six- to twelve-digit numbers, however, the products become rather large for the average calculator.

4-5. THE PECULIARITIES OF REPEATED DIGITS

There are some interesting peculiarities that involve the repetition of similar digits such as 333, 888888, 4444, etc. The oddities are related to the number of digits used, rather than the value of the repeated digits. Starting out with the use of two similar digits, we encounter the following rule: Any double number of similar digits divided by the sum value of the digits equals 5.5. This is illustrated in the following examples:

$$22 \div 4 = 5.5$$
$$66 \div 12 = 5.5$$
$$11 \div 2 = 5.5$$

When three similar-value digits are involved, the division by the sum value of the digits will always equal 37, as shown by the following examples:

$$111 \div 3 = 37$$
$$222 \div 6 = 37$$
$$333 \div 9 = 37$$

The quotient begins to get larger as more numerals are involved. Thus, when we use four similar-value digits the division by the sum value of the numerals produces 277.75 regardless of the numerals selected:

$$4444 \div 16 = 277.75$$
$$7777 \div 28 = 277.75$$
$$9999 \div 36 = 277.75$$

A rather peculiar quotient is obtained when any five similar digits are divided by the sum of the digits:

$$33333 \div 15 = 2222.2$$
$$44444 \div 20 = 2222.2$$
$$88888 \div 40 = 2222.2$$

For a six-numeral group divided by the sum of the numerals, we always obtain a quotient of 18,518.5:

$$222222 \div 12 = 18{,}518.5$$
$$444444 \div 24 = 18{,}518.5$$
$$555555 \div 30 = 18{,}518.5$$

Any seven-digit group of repeated numerals produces a quotient of 158,730.1429 when divided by the sum of the numerals:

$$1111111 \div 7 = 158{,}730.1429$$
$$3333333 \div 21 = 158{,}730.1429$$
$$8888888 \div 56 = 158{,}730.1429$$

Eight repetitive digits divided by their sum produces a quotient of 1,388,888.875 regardless of the value of the individual numerals:

$$11111111 \div 8 = 1{,}388{,}888.875$$
$$33333333 \div 24 = 1{,}388{,}888.875$$
$$77777777 \div 56 = 1{,}388{,}888.875$$

An odd quotient is obtained when any similar digits are formed into a group of 9:

$$222222222 \div 18 = 12345679$$
$$555555555 \div 45 = 12345679$$
$$999999999 \div 81 = 12345679$$

Obviously, in the foregoing examples the selection of 1s to make up the dividend results in a divisor equal to the number of 1s used. Hence, a new rule can be formulated:

Any group of repetitive numerals divided by the sum of the numerals and then multiplied by the number of numerals involved produces a product composed of a series of 1s equal to the number of original numerals.

Thus, using 666 as an example, the sum is 18 and the quotient is 37. Multiplying the latter by 3 = 111. Similarly, 77777 divided by 35 = 2222.2, and when this quotient is multiplied by the number of digits (5), the product is 11111 (see also Sec. 4-1, Recurring Sequence of Ones).

4-6. THE ODD ALTERNATES

An odd sequence of products occurs when the numeral 6 is progressively halved and divided continuously by a succession of 2s. When the digits of the resulting quotients are added (ignoring the decimal point), the sum will always consist of alternate digits of 3, 6, 3, 6, 3, 6, etc.:

$$6 \div 2 = 3$$
$$3 \div 2 = 1.5 \text{ (sum of digits } = 6)$$
$$1.5 \div 2 = 0.75 = 12 = 3$$
$$0.75 \div 2 = 0.375 = 15 = 6$$
$$0.375 \div 2 = 0.1875 = 21 = 3$$
$$0.1875 \div 2 = 0.09375 = 24 = 6$$
$$\text{etc.}$$

4-7. THE MAGIC OF 37037

Using the number 37037 as the multiplicand and the numeral 3 as the multiplier, we obtain 111111 as the product. When the multiplier is increased by another 3, each product digit increases by 1. The same increase by 1 occurs every time the multiplier is increased by 3:

$$37037 \times 3 = 111111$$
$$37037 \times 6 = 222222$$
$$37037 \times 9 = 333333$$
$$37037 \times 12 = 444444$$
$$37037 \times 15 = 555555$$
$$\text{etc.}$$

The products are progressively increased by 1 up to the multiplier of 27, where the product is 999999.

When 3 more is added to the multiplier and the 37037 number is multiplied by the new multiplier of 30, we obtain 1111110. There are now some odd aspects to the new product. The first and last digits will no longer coincide with the center ones, but their sum will.

Thus, when the preceding 30 multiplier is used, the 1 digit at the extreme left plus the 0 to the right = 1; this strange situation is now repeated for additional increases of 3 in the multiplier:

$$37037 \times 30 = 1111110$$

$$37037 \times 33 = 1222221$$

$$37037 \times 36 = 1333332$$

$$37037 \times 39 = 1444443$$

The pattern for the foregoing products, wherein the first place plus the last place digits have a sum equal to the value of each center digit, holds true for a multiplier ranging up to and including 135, as shown by the following example starting with a multiplier of 126:

$$37037 \times 126 = 4666662$$

$$37037 \times 129 = 4777773$$

$$37037 \times 132 = 4888884$$

$$37037 \times 135 = 4999995$$

After the multiplier of 135 there is still a pattern similar to the previous examples, although now somewhat more complex. Multiplying 37037 by 138 produces 5111106. Note, however, that when the last place 5 and the first place 6 are added, we get 11 and these digits still coincide with the center ones! For multipliers above 138 the pattern becomes more complicated as, for instance, $37037 \times 141 = 5222217$. Now the first place digit of 7 added to the last place digit of 5 gives a sum of 12, and the second place 1 in the original product is added to the second place 1 of the sum to produce dual 2s. Similarly, $37037 \times 144 = 5333328$, and adding the left and right numerals gives us $8 + 5 = 13$. Adding the second place digit of the original product and the sum we obtain 33, again coinciding with the center numerals.

In the higher-range multipliers, the pattern again is an odd one. For a multiplier of 282, for instance, we obtain: $37037 \times 282 = 10444434$. Now the last place numeral 1 is added to the second place 3 to produce again the same digits as the center ones. Adding 3 to the multiplier yields $37037 \times 285 = 10555545$, thus continuing with the new pattern. Similarly, $37037 \times 303 = 11222211$, wherein the first two digits are added to the last two digits to produce coinciding 2s. Also, $37037 \times 306 = 11333322$, and each of the first place digits

Sec. 4-8   Let's Look at Quotients

is added to the last place digits to produce coinciding 3s. This pattern continues for a long string of multipliers that are increased by 3 each time.

4-8. LET'S LOOK AT QUOTIENTS

Some peculiar mathematical results are obtained for specific dividends. The divisors need not be out of the ordinary to produce the odd quotients. One example is having a dividend wherein 1 starts at the left, followed by two 2s, three 3s, etc. Dividing by 5 yields:

$$1,223,334,444 \div 5 = 244,666,888.8$$

Note the quotient still starts out with a single digit (now a 2), followed by two repeated 4s, three 6s, etc. Now, however, the digit values increase in increments of 2 rather than 1 as with the dividend. Another odd result is obtained by dividing the descending digits number 987,654,321 by 8. Oddly, we obtain almost a perfect ascension of numerical values except for the missing 8 and the fractional section:

$$987,654,321 \div 8 = 123456790.125$$

Curiously, however, if we transpose the 1 and 2 in the original dividend we obtain a perfect numerical ascension without any missing numerals or remaining fractions:

$$987,654,312 \div 8 = 123,456,789$$

When the number 1,001,001 is used as the dividend some strange quotients occur for certain divisors as illustrated by the following:

$$1,001,001 \div 8 = 125,125.125$$
$$1,001,001 \div 5 = 200,200.200$$
$$1,001,001 \div 4 = 250,250.250$$
$$1,001,001 \div 2 = 500,500.500$$

Dividing by 9 also produces an odd quotient, although now we run into four decimal places:

$$1,001,001 \div 9 = 111,222.3333$$

4-9. THE DIGIT'S SUM EQUALS THE CUBE ROOT

There are many instances where the sum of the digits in a number equals the cube root. The sum of the digits in the number 19683, for instance, is 27. The latter number is also the cube root of 19683. Two more examples follow, although there are numerous other such oddities:

The sum of the digits in 17576 is 26

The cube root of 17576 is 26

The sum of the digits in 5832 is 18

The cube root of 5832 is 18

There are also many with reverse digit answers, wherein the sum of the digits has the same numerals as the cube root, but in reverse order:

Digits sum of $148877 = 53$

$\sqrt[3]{148877} = 35$

Digits sum of $238{,}328 = 26$

$\sqrt[3]{238{,}328} = 62$

4-10. WHY ALWAYS 22332233?

One could, of course, find numerous combinations of multiplicands and multipliers to produce a product of 22332233. Strangely, however, the same product of repetitive 22s and 33s is obtained for many multiplicands also consisting of two or more repetitive numbers. A few examples are:

$5 \times 4466446.6 = 22332233$

$7 \times 3190319.0 = 22332233$

$11 \times 2030203.0 = 22332233$

$14 \times 1595159.5 = 22332233$

$22 \times 1015101.5 = 22332233$

$25 \times 893289.32 = 22332233$

$28 \times 797579.75 = 22332233$

etc.

4-11. THE NUMERAL 8 IS STRANGE

The oddities of 9s were detailed in Sec. 4-3, and one example showed that when a 9 is either the multiplier or the multiplicand the final sum of digits in the product is always 9 (2 × 9 = 18 and when the 1 and 8 are added we end up with 9). The numeral 8, however, also has a strange oddity as exemplified by the partial listing that follows. Note that the final digit sums of products starting with 1 × 8 = 8 decrease in value for successive multipliers. After 1 is reached (at 8 × 8 = 64 = 10 = 1), the descending values start again with 9. Thus, as the multiplier value or count increases, the digit sum again decreases to 1 at the 17 × 8 = 136. At 18 × 8 the descending count is repeated, and this pattern would continue indefinitely for multipliers of increasing values.

$$
\begin{aligned}
1 \times 8 &= = 8 8 \\
2 \times 8 &= 16 = 7 7 \\
3 \times 8 &= 24 = 6 6 \\
4 \times 8 &= 32 = 5 5 \\
5 \times 8 &= 40 = 4 4 \\
6 \times 8 &= 48 = 12 = 3 \\
7 \times 8 &= 56 = 11 = 2 \\
8 \times 8 &= 64 = 10 = 1 \\
9 \times 8 &= 72 = 9 9 \\
10 \times 8 &= 80 = 8 8 \\
11 \times 8 &= 88 = 16 = 7 \\
12 \times 8 &= 96 = 15 = 6 \\
13 \times 8 &= 104 = 5 \\
14 \times 8 &= 112 = 4 \\
15 \times 8 &= 120 = 3 \\
16 \times 8 &= 128 = 11 = 2 \\
17 \times 8 &= 136 = 10 = 1 \\
18 \times 8 &= 144 = 9 9 \\
19 \times 8 &= 152 = 8 8 \\
20 \times 8 &= 160 = 7 7
\end{aligned}
$$

etc.

4-12. DUPLICATES ARE REPETITIVE

In the following mathematical process the resultant product, sum, remainder, quotient, and square root are all composed of numbers having duplicate digit groups. Initially the number 21212 is multiplied by 3. Next add 12121 to the product obtained. Then subtract 52525 from the sum and divide the remainder by 3. Finally, take the square root of the quotient. Proof:

$$
\begin{array}{r}
21212 \times 3 = 63636 \\
+\,12121 \\
\hline
75757 \\
-\,52525 \\
\hline
23232
\end{array}
$$

and the division by 3 yields 7744
The square root of $7744 = 88$.

Similarly, the process could, of course, be undertaken in reverse with these same duplicate digit groups appearing. Thus, 88 is squared, the product multiplied by 3, and the number 52525 added. Next 12121 is subtracted and the remainder divided by 3 to yield 21212.

4-13. FOLDING LIMIT IS EIGHT

Take any square sheet of paper and regardless of how thin it is it will be impossible to fold it by hand more than eight times. In many instances seven folds is a virtual limit, because the eighth no longer resembles a neat, flat fold.

4-14. ODD, STRANGE, AND COMPLICATED

Take the number 3333333.3 and subtract it from 2800000.5. If you now multiply the remainder by 5 and divide by 6 you get 444444. So far we only have repetitive 3s as a start and repetitive 4s as a final quotient. Although this may appear odd, it becomes more strange if we take an additional step. Take 4444444.4 and subtract 3911111.6 from it. Then multiply the remainder by 5 again and divide by 6, and we get the same quotient as before: 444444. The strange thing is that if we take the two subtrahends and subtract the smaller from the larger the remainder is a series of 1s:

$$3911111.6$$
$$-2800000.5$$
$$\overline{1111111.1}$$

If we increase the original number to 5555555.5 and subtract 5022222.7 and multiply the remainder by 5 and divide by 6 again as before, we get the same quotient as before: 444444. Again, if we subtract one subtrahend from the other we again get a series of 1s:

$$5022222.7$$
$$-3911111.6$$
$$\overline{1111111.1}$$

4-15. IT BECOMES ITS OWN $1/X$

The number 1.6180339887 has been called *phi* or *tau* in the literature and has some peculiar characteristics. It is the only positive number that becomes its own reciprocal when we subtract 1 from it. Actually, the number itself is an approximation, and in some references it is given as 1.618034.

In equation form our original number of 1.618033987 is formed by the equation:

$$\frac{1 + \sqrt{5.}}{2}$$

This equation has found some usage in geometrics and some applications in electronics and architecture.

4-16. IT PROVES NOTHING

The following equations are concerned with a series of 3s and a divisor of 6 which are oddly interrelated. Although the results do not prove much, the problem structures are of some academic interest. To do these equations, however, will require a ten-digit calculator. The first three equations have dividends of 20, 23, and the reverse of the latter, 32. Each produces an identical quotient, but when the quotient, in turn, is used as a dividend, we obtain a series of 5s as shown in the following:

$20 \div 6 = 3.333333333$

$23 \div 6 - 0.5 = 3.333333333$

$32 \div 6 - 2 = 3.333333333$

But $3.333333333 \div 6 = 0.555555555$

4-17. REPETITIVE RECIPROCALS

The reciprocal function is usually indicated on a calculator as $1/x$, representing division of 1 by the number in question. Curiously, some reciprocals result in a series of repetitive numerals, some representing a series string that would stretch out to infinity. Note the perfect sequential progressions in values shown initially.

$1/9 = 0.11111111 \ldots\ldots n$

$1/4.5 = 0.22222222 \ldots\ldots n$

$1/3 = 0.33333333 \ldots\ldots n$

$1/2.25 = 0.44444444 \ldots\ldots n$

$1/1.8 = 0.55555555 \ldots\ldots n$

$1.15 = 0.66666666 \ldots\ldots n$

The progression shown ends here unless a large amount of numerals are used such as $1/1.285714287$. For a series of 8s, however, $1/1.125$ yields $0.88888888 \ldots\ldots n$. Odd results are also obtained with other numbers. For instance, $1/1.21212 = 0.825000825$. Note the following consequence of adding fractional-value 3s to the denominator:

$1/3.3 = 3030303030$, etc.

$1/3.33 = 3003003003$

$1/3.333 = 300030003000$

4-18. THE 9s STAND ON THEIR HEADS

Divide any groups of 9s (such as 9999) by 1.5 and thus set them upside down, as 6666. This process functions regardless of how many 9s are used, as, for instance, dividing 99999999 by 1.5 yields 66666666. Similarly, we can tumble 6s by multiplying them by 1.5, as $666666 \times 1.5 = 999999$.

With a little manipulation we can obtain similar results. If, for instance, 9696 is divided by 1.6 and 606 added to the result, we obtain 6666. By changing the number to be added to 909 instead of 606 we reverse the original order of 9696 to 6969, thus tumbling both the 6s and 9s of the original 9696 number. By carrying the basic idea a little further we can tumble the 6s and retain the 1s in 16161616 if we multiply first by 6 then subtract from the product the number 5050505 to give us a final number of 91919191.

If three 6s are used, and multiplied by any number, the final answer will always yield a 9 when the digits are added. Thus $666 \times 23 = 15318$ and $1+5+3+1+8 = 18 = 9$. Thus, we've taken the three 6s and converted them to a single digit (9), which is the 6 upside down again. (This is related to the strange oddity of 9s discussed in Sec. 4-3.)

4-19. REVERSE AND SUBTRACT

Take any sequence of three numerals, reverse the order of sequence and subtract the smaller from the larger. Now add 801 and the answer will always be 999. (Note the digits in 801 add to 9.) As an example, the number 543 in reverse is 345 and when the latter is subtracted from the former we get 198. Adding 801 produces 999.

This little oddity is based on the fact that the same remainder is obtained for any such process involving three consecutive numerals. Actually, this process applied to two-digit numbers always produces a 9; three-digit numbers always produce 198, and four-digit numbers always produce 3087, and so on. Curiously, when adding the digits of any remainder we always end up with a 9. Thus, for a three-digit remainder of 198, the sum of the digits is a 9 as is also the sum of the digits (3087) obtained with the four-digit number. Several examples are shown:

$32 - 23 = 9$ $\qquad$ $321 - 123 = 198$
$54 - 45 = 9$ $\qquad$ $654 - 456 = 198$
$98 - 89 = 9$ $\qquad$ $987 - 789 = 198$
$\qquad\qquad\quad 4321 - 1234 = 3087$
$\qquad\qquad\quad 6543 - 3456 = 3087$
$\qquad\qquad\quad 5432 - 2345 = 3087$

Another version of the foregoing principle is the use of numbers such as 545, 767, and so on. Thus the center numeral is one lower than the left and right numerals. If we reverse the orders of these numbers and subtract we always get 91 for an answer as shown by the three examples below:

$$\begin{array}{ccc} 989 & 212 & 545 \\ -898 & -121 & -454 \\ \hline 91 & 91 & 91 \end{array}$$

If the central numeral is two less than the outside numerals, the subtraction process always yields 182 as shown in the following examples:

$$\begin{array}{ccc} 313 & 646 & 979 \\ -131 & -464 & -797 \\ \hline 182 & 182 & 182 \end{array}$$

If the central numeral is three less than the outside numerals, we will always obtain 273 for the answer:

$$\begin{array}{ccc} 414 & 969 & 636 \\ -141 & -696 & -363 \\ \hline 273 & 273 & 273 \end{array}$$

Note in the first example (where the answer was 91) the sum of the digits equals 10. When the central numeral was two less, the answer was 182 and the sum of the numerals equals 11. When there was a three-numeral difference between the center and outside numerals, the sum of the numerals (273) equals 12. Thus we have a progressive increase in the value of the numerals in the answer for an increasing difference between the center and outside numerals.

5
THE MAGIC NUMBER SEQUENCE OF FIBONACCI AND INFINITY

5-1. THE BASIC SEQUENCE

It is interesting that many of our mathematical curiosities have been with us for many years, preserved in the literature and, since their origin, stimulating the minds of countless persons of all ages. One such oddity is the number sequence formulated by Leonardo of Pisa, commonly known as Fibonacci. This famous Italian mathematician of the thirteenth century published his *Liber Abaci* in 1202, which became a noteworthy foundation for subsequent mathematical books in succeeding years. In addition, the book was instrumental in introducing Arabic notation in Europe.

The Fibonacci number sequence is a most curious item, and the ramifications of it are most interesting. In addition, it has some very practical applications, as shown later. First, however, let us examine the structure of the number sequence. If we set down a 1 and a 2 and add them, we get 3 for the start of the series. Then we add each last number to the preceding one, to get a new number, as $1 + 2 = 3$, and $2 + 3 = 5$, etc. Continuing in this fashion, the series becomes:

$$
\begin{array}{c}
1 \\
2 \\
3 \\
5 \\
8 \\
13 \\
21 \\
34 \\
55 \\
89 \\
144 \\
233 \\
377 \\
610 \\
987 \\
1597 \\
2584 \\
\text{etc.}
\end{array}
$$

5-2. PRODUCT PECULIARITIES

The series itself doesn't look too impressive, and superficially it appears to have no characteristics other than being composed of successive sums of the preceding two numbers. When we analyze some of the hidden aspects of this number sequence, however, we are in for a few surprises. If, for instance, you square any two successive numbers and add the products thus obtained, you get another number within the series. Selecting the numeral 2, for instance, yields 4 (2×2), and squaring the next numeral (3) gives us 9. Adding the two products: $4 + 9 = 13$, the sixth number in the sequence. Similarly, $3 \times 3 = 9$, and $5 \times 5 = 25$; then, $9 + 25 = 34$, the eighth number in the series. The same end results prevail, regardless of the extent of the sequence. If you feel like wrestling with large numbers, you can prove this peculiarity as far along the number sequence as you wish. For instance, 55^2 plus $89^2 = 10,946$ (3025 + 7921). Now you are reaching to the twentieth number of the sequence as you can prove by expanding the sequence.

Because of the symmetry of this number series, it follows that the answers obtained from the foregoing process will be an orderly set of numbers having a definite pattern related to the series itself. Hence, you will find that the answers are always separated by two numbers in the series:

$1^2 + 2^2 = 5$ (the fourth number of the series)
$2^2 + 3^2 = 13$ (the sixth number of the series)
$3^2 + 5^2 = 34$ (the eighth number of the series)
$5^2 + 8^2 = 89$ (the tenth number of the series)
$8^2 + 13^2 = 233$ (the twelfth number of the series)

etc.

Another (of many) interesting fact about this series is that the product of any two numbers (separated by two numbers) always differs from the product of the intervening numbers by 1. Suppose, for instance, that we select 2 and 8 to form our first product. (The *in-between* numbers will then be 3 and 5.) The product of 2 and 8 is 16, and the product of 3 and 5 is 15. If you select 3 and 13, the product will be 39. Again, the intervening numbers (5 and 8) will give a product (40), which differs from the product formed by the two separated numbers by 1. As an additional example, $144 \times 610 = 87,840$ and the product of the two numbers in between is given by: $233 \times 377 = 87,841$.

5-3. THE MAGIC MULTIPLIER

Another interesting aspect of this series is the magical qualities of 11 when used to form products. If, for instance, you multiply the sixth number by 11 and add the first number to the product thus obtained, you will have the eleventh number in the sequence! Proof: $11 \times 13 = 143$. Adding the first number (1) to the product gives us 144, the eleventh number. Similarly, if you multiply the seventh number by 11 and add the *second* number, you will obtain the twelfth number of the sequence: $11 \times 21 = 231$. Adding the second number (2) gives us 233, the twelfth number. Multiplying the eighth number by 11 and adding the third number will give us the thirteenth number, etc.

This curious property of the number 11 as a multiplier holds even though other numerals are used to start the series. Assume we select 3 and 6 as our initial numbers. A new Fibonacci series will be formed having a number sequence differing from the one shown earlier, which started with 1 and 2:

$$3$$
$$6$$
$$9$$
$$15$$
$$24$$
$$39$$
$$63$$
$$102$$
$$165$$
$$267$$
$$432$$
$$699$$
$$1131$$
$$1830$$

etc.

Here, again, if you multiply the sixth number by 11 and add the first number, you will obtain the eleventh number: $39 \times 11 = 429 + 3$ (the first number) $= 432$ (the eleventh number). Similarly, if we multiply the seventh number (63) by 11 we obtain 693, and when we add the second number (6) we get the twelfth number, 699! Again, if we multiply the eighth number by 11 and add the third number, we will get the thirteenth number, etc.

The first number of the series does not have to be of a value numerically less than the second number. We could have reversed the 3 and 6 used initially to form the preceding series and the unusual characteristics of the 11 multiplier would still hold:

```
  6
  3
  9
 12
 21
 33   (33 × 11 = 363 + 6 = 369—the eleventh number)
 54
 87
141
228
369
597
```
etc.

It makes no difference what the first two numbers are, the magical powers of the 11 multiplier still prevail. The first number could be ridiculously high, or even astronomical compared to the second number, and all rules still hold. The following shows the sequence formed when the first number contains five places and the second number only has two places:

```
  78563
     24
  78587
  78611
 157198
 235809 × 11 =   2593899
 393007        +   78563   (the first number)
 628816        ─────────
1021823        =  2672462  (the eleventh number)
1650639
2672462
4323101
```

5-4. THE MYSTIC SPIRALS

Is the number series of Fibonacci an area of pure mathematics that, although fascinating, is unrelated to practicality and reality? Not at all. One truly mysterious aspect of the number sequence is its relationship to nature's arrangement of florets in daisies and sunflower core discs, pine cone scales, and other plant growths with spiral patterns in leaf growth. An inspection of the core of the daisy reveals tiny florets so grouped that they form spirals. Two spiral groupings are in evidence, one with a clockwise configuration and the other counterclockwise. Each group contains a given number of spirals, and for the daisy blossom the usual numbers are 21 and 34 for the clockwise and counterclockwise sets. Note that these are two sequential numbers of the Fibonacci series first given: 1, 2, 3, 5, 8, 13, 21, 34, 55, 89, etc. Similarly, the opposing spirals in a sunflower may number 34 and 55, again a part of the Fibonacci series. If a greater number of spirals is present, they will still conform to the number sequence of the series.

The scales in a pine cone are also in a spiral formation, with 5 in one direction and 8 in another, again displaying a mysterious relationship to the Fibonacci series. Similarly, the pineapple exhibits its bump arrangement with a number arrangement of 8 and 13, again part of the series!

Scientists, botanists, and others have attempted explanations of the mysterious relationships between the spirals (which are also related to the area of logarithm mathematics, since they are logarithmic spirals) and the Fibonacci series. Some theories relate the energy that contributes to plant growth to the precise mathematical rules of electrical energy. Some attribute the mathematical relationship to the divine proportions that have been recognized for centuries, and which are covered in greater detail in the section that follows.

5-5. THE GOLDEN RATIO

For many centuries certain ratios and proportions have been used in architecture, sculpture, and painting that have proved to have the highest degree of appeal to the viewer and that were considered to be the ultimate in artistic endeavors. Because such ratios and proportions in artistic physical dimensions were instinctively recognized as the epitome of delineation, the term *divine* was applied to them. Because of the Golden Age of art, the word *golden* was also used as an adjective when describing such dimensions. Hence, the phrases *divine proportions*, *golden proportions*, and *golden ratios* are still in vogue for describing specific physical dimensions in art. Again we find an uncanny tie-in with the Fibonacci number series when we inspect certain ratios in it.

Analytical studies of the significant dimensions of ancient Greek art indicate a particular preference for proportions having ratios related to $\sqrt{2}$, $\sqrt{3}$, etc., with a strong leaning to $\sqrt{5}$. Thus, a painting with ratios involving such square roots, or an architectural design based on it, proves to have the greatest aesthetic appeal to the average viewer. The rectangle thus formed has, in consequence, been termed the *golden rectangle*, and the ratio of one side to the side adjacent, the *golden ratio*.

The equation for the golden ratio is:

$$x = \frac{1 + \sqrt{5}}{2}$$

Since the square root of 5 is approximately 2.2361, the equation becomes:

$$x = \frac{1 + 2.2361}{2} = \frac{3.2361}{2} = 1.618, \text{approximately}$$

Thus, the golden rectangle has dimensional proportions of 1 to 1.618.

How is the golden ratio related to the Fibonacci number sequence? Consider again the initial sequence given: 1, 2, 3, 5, 8, 13, 21, 34, etc. If we multiply any term in the series by the x value of the golden ratio, we get a close approximation of the next number. For instance, 5 × 1.618 = 8.090, which differs from the next number (8) of the series by only 0.09. If we multiply 21 by 1.618 we get 33.978, which is only 0.022 removed from the next number of the series, 34. As higher numbers of the series are multiplied by the 1.618 number, we get closer and closer to the exact value of the next number in the series.

Consider yet another curious relationship of the Fibonacci series to the golden ratio. For the higher numbers of the series, the ratio of any sequential pair of numbers is approximately equal to the golden ratio number 0.618. Thus,

$$55 / 89 = 0.618+$$
$$89 / 144 = 0.618+$$
$$144 / 233 = 0.618+$$
$$233 / 377 = 0.618+$$
$$377 / 610 = 0.618+$$
$$610 / 987 = 0.618+$$

etc.

Thus the famous Fibonacci number sequence presents us with a number of mysteries that have intrigued not only the mathematician but the artist and the psychologist for many years, Perhaps, in time, the Fibonacci series will prove to have other meaningful relationships, as yet unrecognized, but somehow related to some practical phase of human activity.

5-6. THE INFINITY CONTRADICTION

Anyone having had some math courses has become familiar with the traditional infinity concept. When we think of infinity with respect to time we think of an endless sequence of days. Mathematical infinity can relate to both negative and positive numbers. We can increase expressed numerical values without any end since any given number value can be increased at will. Similarly, numbers with negative values can also be increased in magnitude indefinitely and without restrictions. The infinity concept has found practical utilization in the calculus and was given new status by Albert Einstein. With the application of this great mind to universal phenomenon, science came of age, and to the surprise of many, the older laws of Newton were superseded by startling new concepts. Many mysteries remained, however, and it must be left to some future genius to make additional contributions to our store of knowledge.

The infinity concept can provide some strange situations in seeming contradiction to established rules. Mathematically, if we take successive halves of any number there will always be one half left. Obviously, we eventually reach such fractional values as one billionth, but regardless of how small a value we attain we can always halve the amount. Similarly, if we keep slicing intervals of time we reach the microsecond level and even the much shorter intervals of time in nanoseconds now commonly utilized in electronics. Again, regardless of how small the interval of time, we can conceive of reducing it additionally. Similarly, if we were reducing a given length we would eventually reach space measured in micrometers or smaller, again with no apparent mathematical end since we can always cut whatever we have left in half again indefinitely.

The preceding basic concepts appear valid and foolproof since there doesn't appear to be any valid argument against the fact that mathematically we can always reduce a given quantity by half. Assume, however, that we use a length of one third meter as a reference and think of progressively halving the span by increments of one half. Mathematically, we would always have one half left and would never reach the end. If, however, we take a pencil and move it along this span we will obviously not only reach the end but will move beyond it. Now we have mysteriously spanned all the infinite successive halves and not only reached the infinite end but progressed beyond it. How can we explain this seeming contradiction?

Although several attempts have been made to unfold the mysteries of this situation, it is questionable that the final satisfactory solution is imminent. In considering this infinity-related puzzle, the limitations of the human mind tend to obscure certain fundamental aspects. If we think of cutting a given length in half in successive steps, we tend to allocate, subconsciously, the same interval of time to each successive reduction in length. Thus, when we think of the initial cut we mentally assume it is done in a discrete time interval. After we halve the total length and then perceive cutting the remainder in half, we assume a time interval coinciding with the initial. Thus, in essence, for each cut the halves become progressively shorter but seemingly at the same rate of cutting time that prevailed for the initial halving of the original amount.

Consider, however, what actually occurs when we move a pencil or a pointer along a given length. We progressively reach successive half points, but because the pointer is moving at a fixed rate of speed the time intervals become shorter for successive halves, and the time consumed to span successive halves decreases proportionally. Because the time interval for each successive span decreases by half, it is as though the progression over successive halves doubles in speed. As the speed for successive halves doubles, we hasten the approach to the end of the span as the successive halves become increasingly shorter in length. It is as though the shorter intervals of length accompanied by increased speed as each remaining half is spanned together permit us to reach finally a zero point in spanning successive halves. Mathematically, however, we cannot prove this since, regardless of the fractional value left after spanning successive halves, there should always be a half left.

The great mathematician George Cantor (1845-1918) delved deeply into the theoretical aspects of infinity and came up with some interesting theorems. In 1882, Cantor disclosed a novel number concept: a process termed one-to-one correspondence. In his proposal the existence of a one-to-one correspondence between the elements of a double string of numbers when the aggregate of one series matched the other is reason to consider them identical. If we construct a row of positive integers to form a row of 1, 2, 3, 4, and so on, logic tells us that each integer must be followed by another in an

Sec. 5-6   The Infinity Contradiction   69

infinite progression. If, however, we match each integer with its numerical double we obtain the following series:

$$1 \ 2 \ 3 \ 4 \ \ 5 \ \ 6 \ldots n$$
$$2 \ 4 \ 6 \ 8 \ 10 \ 12 \ldots 2n$$

Now we have a curious situation, because the number of integers matches exactly the number of even integers, even though the latter are actually a part of the former. A similar situation prevails for a dual series of even and odd numbers, as:

$$1 \ 2 \ 3 \ 4 \ 5 \ \ 6 \ \ 7 \ldots n$$
$$1 \ 3 \ 5 \ 7 \ 9 \ 11 \ 13 \ldots n$$

The validity of the double number series theory is questionable. As soon as you consider a cutoff point in the progression of even and odd numbers the concept of infinity is meaningless since infinity has no cutoff—no end. Even the term *progressing toward infinity* is incorrect, because we can't make any significant advance toward something endless. Even a megameter movement toward infinity is no nearer to an end that doesn't exist than a millimeter movement.

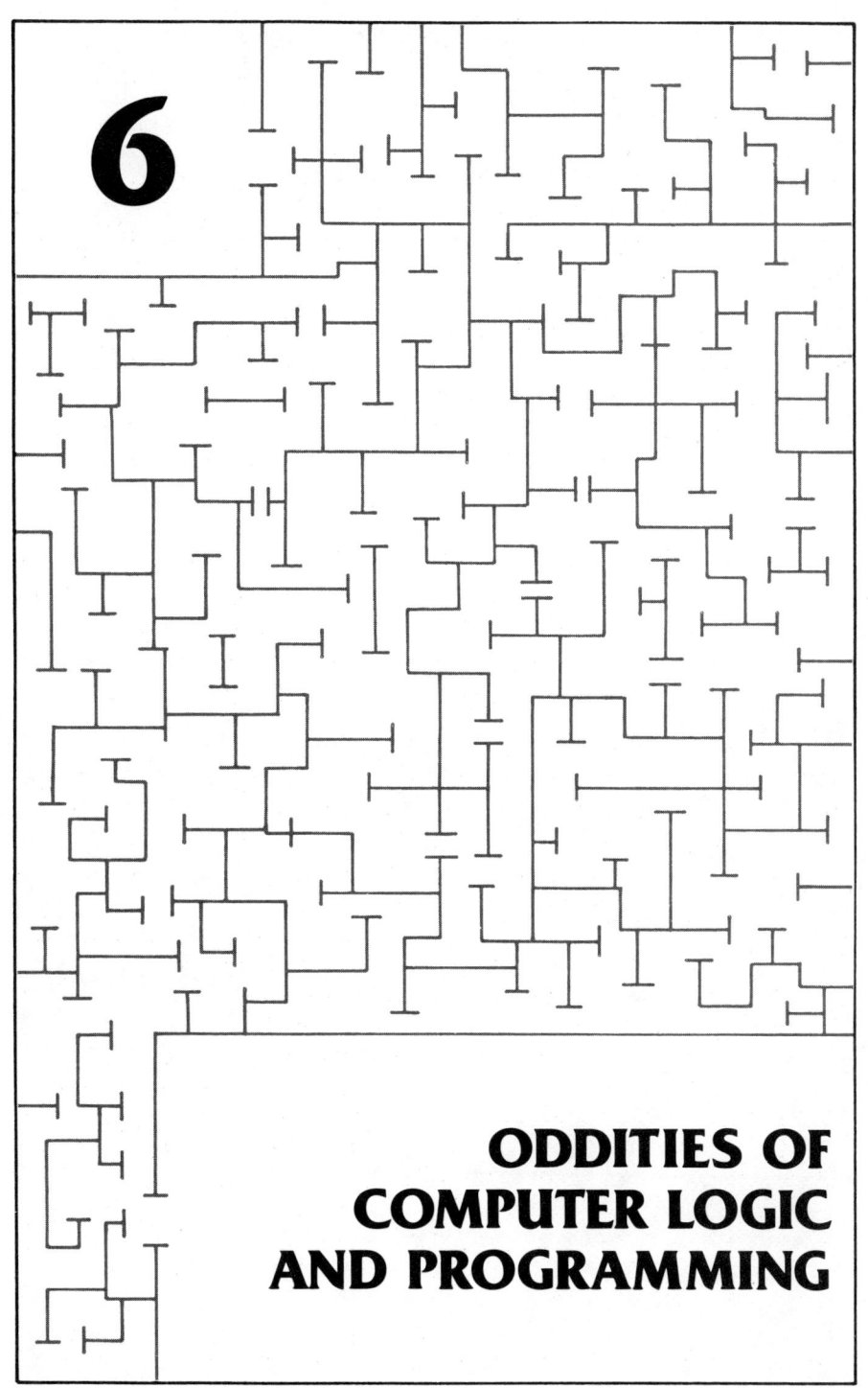

6

ODDITIES OF COMPUTER LOGIC AND PROGRAMMING

INTRODUCTION

With the advent of calculators and computers, we have accepted as routine the ease and speed with which mathematical processes are performed in calculators and the facility with which a computer handles data processing and other complex programs. To an avid mathematician, however, the methodology by which calculators and computers perform their myriad functions creates a sense of surprise and wonder. It cannot be otherwise since we are confronted with a device that performs all the arithmetic functions in existence not only at incredible speeds but by use of an unusual mathematical radix system. Most mathematicians are aware of the radix-2 system in basic form, but until recently, few would have believed that the radix-2 system is as versatile as it has proved to be.

In this chapter we not only explore some of the aspects of the base-2 system but also of the associated science of logic which, surprisingly, had been considered in some depth by certain mathematicians many years before the advent of computers. The discussions of these topics in this chapter also include some of the logic algebra that is applicable and the fundamental diagrams that are relevant.

6-1. LOGIC AND MATH

The entire operational capabilities of all digital computers are closely linked with the laws of logic and the principles of mathematics. One of the oddities is the use of the base-2 arithmetic system (radix-2) within the computer circuitry. To perform all arithmetic functions this means utilizing only two numerals, a 1 and a 0. To those unfamiliar with the base-2 binary system, it would seem impossible that the basic arithmetic functions (as well as complex equations) can be executed utilizing only a series of 1s and 0s. Some aspects and oddities of this number system are included in this chapter.

Of interest, also is the necessity for applying laws of logic to the routing, comparing, and storing of numerical values. Curiously, several early mathematicians and logicians contributed materially to an understanding of logical fundamentals and thus laid a foundation for use of an orderly process in design. One such individual was Charles

L. Dodgson (1832-1898), the English mathematician and author. Dodgson was a noted mathematical lecturer and wrote numerous mathematics texts, including several on geometry and trigonometry. He also described logic processes and theory in two publications: *The Games of Logic* and *Symbolic Logic*. Of considerable interest, however, is the fact that Dodgson used the pen name Lewis Carroll. It was under the latter pseudonym that he wrote a series of children's books that included *Alice's Adventures in Wonderland*, which ultimately became a world classic. Eventually he wrote other such texts including *Alice Through the Looking Glass*. Dodgson was inordinately proud of his mathematical books but less so of his fictional successes and often disclaimed their authorship when queried about them.

Subsequent mathematicians and scholars who contributed theory and rules of logic to the literature included George Boole (1815-1864). Boole published mathematical rules and statements involving logical conclusions. His rules of logic processes are now referred to as *Boolean algebra*. These rules are standard educational materials that are part of every modern course in computer or switching logic. Many others also contributed to some degree to the accumulated knowledge available at present. Symbolic logic came of age when Claude E. Shannon published his thesis for his Master of Science degree at MIT. His thesis was so significant and so important in the area of symbolic logic that he gained immediate recognition. An abstract of the thesis was published in the trade journal of the *American Institute of Electrical Engineers* in 1938 entitled "A Symbolic Analysis of Relay and Switching Circuits." Shannon's brilliant discourse unified the scattered bits of logic information that had been extant and thus permitted widespread practical application of Boolean algebra to switching and gating principles. This logic algebra was not only used in the design of computers but also telephone switching systems and other areas requiring symbolic logic design. Although an indepth treatment of symbolic logic and the binary system would involve several texts, some significant aspects are touched upon in this chapter to illustrate interesting curiosities and oddities that are involved.

6-2. THE PECULIAR BINARY SYSTEM

In computers (as in calculators) accuracy is of prime importance, and to assure reliability the machine's numbering system is confined to only two states. These two states represent the numerals 1 and 0 and

Sec. 6-2 The Peculiar Binary System

are achieved within the computer by assigning different electric potentials to each. Thus if a 0 is represented by a signal having a negative polarity and the 1 by a signal having a positive polarity, there is a sufficient difference to assure an absolute minimum of errors. Even if a single polarity is used, one signal is given a sufficient amplitude difference to prevent identity mistakes during computations.

The reliability achieved, however, extracts a price. The computer does not now perform with our familiar base-10 system but rather in the binary or radix-2. This means that all computations of add, subtract, multiply, divide, square roots, and so on must be performed by arithmetic notations involving combinations of 1 and 0 only. Although only 1 and 0 are used, the same general principles relating to arithmetical place prevail. The binary system is of primary interest to the design engineer or the trouble-shooting technician rather than the programmer. For convenience in human dialog with the computer circuitry, specific programming methods have been developed enabling the computer to accept base-10 numbers and equations and automatically convert them to the binary system utilized in its internal circuits. Since the binary system is unique in its functions in a computer, it is interesting to explore the system additionally in order to observe some of the mathematical oddities that prevail. We will also thus survey the clever manner in which the binary system is manipulated for the convenience of the programmer to facilitate the entering of base-10 data.

As with our base-10 arithmetic notation, the binary system also has specific values for the position of each numeral. For instance, in our base-10 system a numeral at first place (to the left of the decimal) is termed the *units position*, as shown in Fig. 6-1(a). In second place we have the *tens* position, in third place the *hundreds*, in fourth place the *thousands*, in fifth place *tens* of *thousands*, and

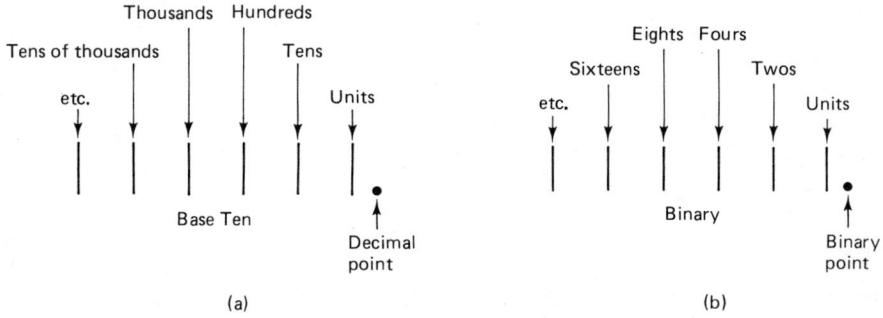

Figure 6-1

so on. Similarly in the binary system, the first place to the left of the decimal (actually the *binary* point) is the *units* position as shown in (b). In second place, however, we now have the *twos* position, in third place the *fours* position, in fourth place the *eights* position, and so on. Note that each progressive place to the left increases to twice the value of the preceding place.

In view of the foregoing, we are able to express any value we wish in binary system. If we have the number 1101 the aggregate value of the numerals is 13 as shown in Fig. 6-2(a). If we have the binary number 10111011 as shown in (b), the accumulated value of the numerals equals 187 as shown. As with our base-10 system, a zero has no value but does establish the number of places within a number. Thus any number magnitude can be expressed, and the only disadvantage over the base-10 system is that more numerals are needed to express the same number value. Our 187 in the base-10 system has only three numerals but the equivalent number in binary has eight numerals.

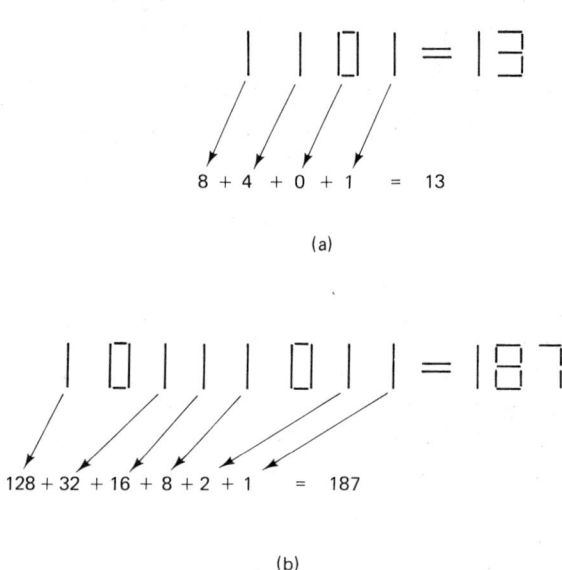

Figure 6-2

A listing of binary numbers in sequence plus their decimal equivalent to a value of 16 is shown in Table 6-1. Such a table is actually easy to construct up to any numerical magnitude because of the particular pattern that prevails. Note that the column of binary numerals in first place alternates 0s and 1s down the column.

Sec. 6-2   The Peculiar Binary System

TABLE 6-1 Binary Numbers
and Base-10 Equivalents

Binary	Base-10
0000	0
0001	1
0010	2
0011	3
0100	4
0101	5
0110	6
0111	7
1000	8
1001	9
1010	10
1011	11
1100	12
1101	13
1010	14
1111	15
10000	16

In second place, however, there are alternate pairs of 0s and 1s. For third place we have a succession of four 0s, four 1s, four 0s, and so on regardless of the length of the column. Note that in the fourth place column we have eight 0s, eight 1s, and so on for the entire column. Successive columns would have groups of sixteen 0s and 1s, the next column a series of thirty-two 0s and 1s, and so on.

A base-10 number can be converted to its binary equivalent by dividing the base-10 number by 2 and writing down the successive numerals in the remainder. The latter sequence forms the binary number that has a value equivalent to the base-10 number. When dividing, the numerals in the remainder will always be either a 1 or a 0. As an example, to ascertain the binary number equivalent of 76 entails the following:

	Quotient	*Remainder*
$\frac{76}{2} =$	38	0
$\frac{38}{2} =$	19	0
$\frac{19}{2} =$	9	1

	Quotient	Remainder
$\frac{9}{2} =$	4	1
$\frac{4}{2} =$	2	0
$\frac{2}{2} =$	1	0
$\frac{1}{2} =$	0	1

Thus, the sequence of numerals in the remainder are written down with the top numeral of the remainder comprising first place. Consequently we obtain the following binary number: 1001100, or 64 + 8 + 4 = 76, the original number.

Since we can express any number of any magnitude, it follows that we can also manipulate the numbers to perform all arithmetic operations. In arithmetic addition, for instance, a binary 101 added to 10 produces a sum of 111 as shown below. As in the base-10 system the carry function is also present. Thus, if 111 is added to 100 as shown next, the addition of 1 and 1 produces a 0 with 1 to carry.

```
  101  (5)  addend        111   (7)
+  10  (2)  augend      + 100   (4)
  ---                    ----
  111  (7)  sum          1011  (11)
```

When the carry function occurs in the earlier places, it has the same progression as in ordinary arithmetic. If we add 88 to 88, the first place addition of 8 + 8 would produce a 6 with 1 to carry. This adds to the numerical value of second place and a carry function extends to third place, and so on. Thus, if we add the binary 101 to 11 as shown in the following, the first place 1s = 0 and carry 1. The 1 is carried to the second place, is added to the second place 1 in the augend, and hence equals 0, with another 1 to carry. The carry unit is added to the third place 1 in the addend again, resulting in 0 carry 1. Now the sum (1000) has a value of 8 in the binary system, which is the correct sum. The addition of three numbers as shown below embraces the same principles. The first place 1s produce a 0 with 1 to carry. Since we already have two 1s in the second place column, which would equal 0 with 1 to carry, the adding of the extra digit results in a 1 in the sum with 1 to carry.

Sec. 6-2   The Peculiar Binary System

```
 101  (5)         11  (3)
  11  (3)         01  (1)
 ————            ———
1000  (8)         10  (2)
                 ———
                 110  (6)
```

During subtraction we again have basic procedures identical to those found in base-10 arithmetic. Subtracting 100 from 111 cancels out the third place 1s, leaving 11 as shown below. When, however, a borrow function is necessary we follow the same procedure as in base-10 as shown by the next example. Subtracting binary 1 from 110 necessitates the borrowing of the second place 1. Since a second place 1 added to a 0 produces 10, which has a value of two, the subtraction of 1 from 10 leaves a 1 in the remainder as shown. Since the second place 1 was borrowed, a 0 appears in second place. The third place 1 in the minuend is brought down as shown for a remainder of 101, which has a base-10 value of 5.

```
  111  (7) minuend        110  (6)
 -100  (4) subtrahend    -  1  (1)
 ————                    ———
   11  (3) remainder      101  (5)
```

The binary multiplication process is simple because we only have a 0 and 1 to contend with rather than our base-10 numerals from 0 through 9. Thus, if we are multiplying the binary 111 by 10, we get a product of 1110 as shown below. Inverting the multiplicand and multiplier to prove the product produces the usual partial products as also shown below with the same product. If we encounter any 1 + 1 situation in the partial products, they produce a 0 with 1 to carry as explained earlier for the addition process.

```
  111  (7) multiplicand      10
X  10  (2) multiplier      X 111
————                       ————
 1110  (14) product          10
                             10   partial
                             10   products
                           ————
                           1110
```

In binary division we follow the same procedures as in the base-10 system. If we divide 1010 (10) by 10 (2), we obtain 101 (5) as shown next. However, if the dividend is 11100 and the divisor is 111, our quotient would be 100 since we must recognize the first and second place 0s as in the base-10 system.

```
            101  quotient                100
divisor  10/1010  dividend           111/11100
            10                           111
           ----                          ----
            10                            00
```

Of interest also is the fact that binary numbers can be expressed as fractional values to the right of the decimal point. In the base-10 system the first numeral to the right of the decimal represents *tenths*, the next *hundredths*, and the next *thousandths*, etc. In the binary system, however, the value sequence to the right of the point is *halves, fourths, eighths, sixteenths*, etc. Thus, in our base-10 system the number 5.3 represents 5-3/10, but in the binary system 101.11 represents 5-3/4. Similarly, 0.1 = 1/2, 0.01 = 1/4, and 0.001 = 1/8 in binary. A few random numbers follow as an additional illustration:

$$101.1 = 5\tfrac{1}{2}$$

$$111.01 = 7\tfrac{1}{4}$$

$$1001.101 = 9\tfrac{5}{8}$$

$$100.1001 = 4\tfrac{9}{16}$$

The foregoing points out a dual weakness present in both the base-10 and binary systems. In the former we must represent three quarters as 0.75 or as the fractional expression 75/100 or 3/4. In binary, however, we can represent three quarters exactly as 0.11 instead of being forced to represent it as a fractional expression. On the other hand, however, in binary we cannot represent twenty-five hundredths exactly but only as 0.01, which represents one quarter. In the base-10 system, however, we get the exact representation by writing 0.25.

6-3. IT'S A REAL COMPLEMENT

In early computer design the computational processes, although performed much more rapidly than could be performed by humans, were actually slow compared to modern computers. The old types used vacuum tubes, large chassis, and separate cabinets with inter-

Sec. 6-3　It's a Real Complement　　　　　　　　　　　　　　　　　　81

connections. The modern usage of chips and high-speed switching transistors provide for extreme rapidity of calculations even when the latter are complex. To expedite the subtraction process in early computers, the *complementing* principle was initiated. This odd procedure performs subtraction by a peculiar additive process. Actually, the complement system could be employed with base-10 numbers and a few examples will help in understanding the binary procedure. The process simply involves the changing of the subtrahend to the representative complement number and adding. For instance, assume that 5246 is to be subtracted from 7964:

$$\begin{array}{r} 7964 \\ -\,5246 \\ \hline 2718 \end{array}$$

To complement, we change the subtrahend so each numeral is the difference between it and 9. If the original number is 5, for instance, it becomes 4; if 2 it becomes 7, and so on. Next we *add* the two numbers:

$$\begin{array}{r} 7964 \\ +\ 4753 \\ \hline 12717 \end{array}$$

The first digit at the left in the remainder is now shifted into the *units* position at the right and added, thus producing the same answer as obtained by the subtraction process:

$$\begin{array}{r} 7964 \\ +\,4753 \\ \hline (1)\ 2717 \\ +\quad 1 \\ \hline 2718 \end{array}$$

The leading digit that was shifted is always the numeral 1 and the shifting process has been termed *end-around carry*. In the base-10 system this complementing process is termed *nines complementing*. The process can also be applied to the binary system, in which case it is termed *ones complementing*. In the binary system the process is much less complex than in the base-10 system. Since only the numerals 0 and 1 are involved, the complementing process simply means substituting one digit for the other. Thus, the entire subtrahend is simply inverted, wherein each 1 becomes a 0 and each 0 becomes a 1. As an example, if the binary 100 is to be subtracted from 1010 we would obtain 110 as shown next. In complementing the subtrahend it becomes 1011 as shown. Performing the end-around carry function places the left-most 1 in the units position and when added to the sum produces 110, the same number as in the subtraction process. In this process it is essential that the subtrahend has as many binary numerals as the minuend. Thus in the example below, the left-most zero is put in place since it will be changed to a 1 as shown.

$$\begin{array}{rl} 1010 & (10) \\ -0100 & (4) \\ \hline 110 & (6) \end{array} \qquad \begin{array}{r} 1010 \\ +1011 \\ \hline (1)\ 0101 \\ +\qquad 1 \\ \hline 110 \end{array}$$

As an additional example assume we wish to subtract 1011 from 10111. The processes showing the normal subtraction method and the ones-complementing system are:

$$\begin{array}{rl} 10111 & (23) \\ -01011 & (11) \\ \hline 1100 & (12) \end{array} \qquad \begin{array}{r} 10111 \\ +10100 \\ \hline (1)\ 01011 \\ +\qquad 1 \\ \hline 1100 \end{array}$$

In computers the process can be simplified considerably. Since the end-around-carry bit is always a 1, it is dropped as a left-most bit and a numeral 1 is always added automatically to the right-most numeral.

6-4. THE DIALOG COMPROMISE

Because the internal circuitry of the computer only uses the binary system, it would be difficult and time consuming if we had to program (enter equations and data) by this system. Consequently, simplified procedures were sought early in computer design. Subsequently shortcut methods were discovered, although the process had to evolve gradually because much depended on advanced computer design and storage capabilities.

Sec. 6-4 The Dialog Compromise

Early programming methods utilized *mnemonic* (aid-to-memory) coding. In this system three or more letters were used to represent the instruction to be performed. Typical was the term LDA, which indicated *load the accumulator*. Addition was simply specified as ADD while manual input was designated by MNI, and so on. Mnemonic coding still had many limitations and subsequently special programs were developed that permitted a more simplified means for dialog with the computer. (The basic aspects of modern program types are covered in Chapter 7.)

With the new statement the programs were easily constructed and could also be read by someone else. In FORTRAN, for instance, an equation statement follows closely the actual structure of a conventional math equation. Thus this particular language has been suitable for solving complex equations and handling numerical data. An equation such as $a+b/c$ would be expressed in FORTRAN as (A+B)/C. Similarly SQRT indicates square root, ALOG 10 indicates logarithm base-10, and so on. Where differences prevail in equation statements, they are necessitated by the limitations of the system. Instead of using the letter x to indicate multiplication, an asterisk is used to avoid confusion with the common x and y symbols used in math. Similarly, to keep numbers on a straight line, a double asterisk is used for exponentiation. Thus (A*B)+C** would mean $(ab)+c^2$.

Other such program languages used identical or similar methods to expedite human dialog with the computer. Special compiler and assembler programs must be utilized to convert high-level languages such as FORTRAN, BASIC, etc., to the machine language. All high-level languages require precise expressions of instruction. If a specific computer command that indicates information transfer is given as GOTO, the substitution of any other words will invalidate the command. If a halt command that causes a computer process to cease is given as PAUSE, the computer will not function with a word entry such as *stop* or *terminate*.

The binary notation itself can be coded to facilitate acceptance of base-10 arithmetical data. One aspect of this is to code binary numbers so they have a form equivalent to base-10 notation. In this system four binary bits are used as a group and held to exactly these four bits throughout. This then permits us to use the four bits to represent a single base-10 numeral. By doing this we can then express a number such as 286 in a binary-coded decimal form as follows:

$$0010 \quad 1000 \quad 0110$$
$$(2) \quad\;\; (8) \quad\;\; (6)$$

Thus, for the first nine digits, pure binary prevails, but after the ninth number the binary-coded system is utilized and numerical data entry into the computer is simplified. Table 6-2 lists some representative decimal numbers and the binary-coded equivalent expressions. Other codes are also utilized in some design practices, but the basic internal machine code is still the pure binary system.

TABLE 6-2 Binary-Coded Decimal Notation

Base-10	Binary-Coded Equivalent		
01	0001		
02	0010		
03	0011		
04	0100		
05	0101		
06	0110		
07	0111		
08	1000		
09	1001		
10	0001	0000	
11	0001	0001	
12	0001	0010	
	etc.		
20	0010	0000	
21	0010	0001	
22	0010	0010	
	etc.		
346	0011	0100	0110
758	0111	0101	1000
921	1001	0010	0001

6-5. THE BASIC SYSTEM

Electronic calculators and computers have many circuits in common, and the basic design aspects between the two are virtually identical. Even in modern calculators, significant strides have been made in producing compact units having a wide range of capabilities. The scientific calculators are capable of expressing numbers with exponents and are programmed to furnish trigonometric ratio displays as required. Many have parenthetical capabilities. An acceptance of six or seven parenthetical expressions in one equation is not unusual. Some programmable calculators are also available that have some limited decision acceptance capabilities. In contrast, the computers

Sec. 6-5   The BASIC System   85

not only perform all the functions of calculators but also have a random access memory (RAM) in addition to the read-only memories (ROM) needed to store standard functions. Computers have the capability of accepting specific programs that utilize the internal circuitry to analyze, compute, sort, store, and retrieve selected data as needed. Decision-type commands are common, wherein the computer is required to branch to a different set of commands when a specific condition prevails. Stored preprogrammed routines are always available to perform routine math or data processing procedures.

Modern electronic science has progressed to the point where a full-function computer is contained within a small typewriter keyboard console. Essentially the basic computer sections are as shown in Fig. 6-3. Input and output systems must be utilized and units such as typewriter keyboards, visual displays, printers, and other external items are termed *peripheral units*. These constitute the *hardware* aspects of a computer as compared to the *software* such as programs, standard routines, etc. As shown in Fig. 6-3, input items include typewriter keyboard, external memory units, punch-card readers (if used), as well as linkages from remote input devices. All the input is applied to the central processor unit (CPU), which is the heart of the computer. It is the CPU that contains the various storage facilities, arithmetic units, routing and gating circuits, and other essential linkages. The output from the CPU is fed to a display unit (such as a television screen), electric typewriter, external storage, high-speed printer, or remotely linked output device.

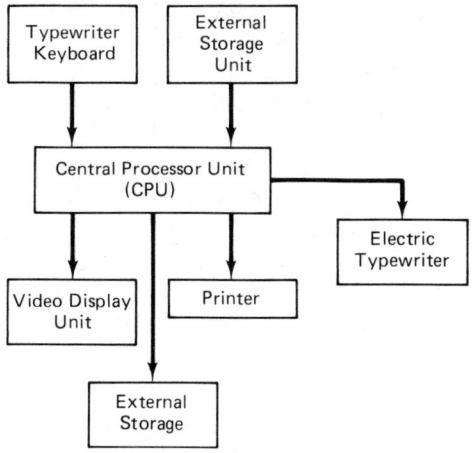

Figure 6-3

6-6. SOME LOGIC SYSTEM ASPECTS

Logic concepts and notations have been taught in colleges for many years and as mentioned in the beginning of this chapter, early mathematicians became pioneers in the exploration of this field. Traditional concepts still prevail, although in the last few decades there have been significant changes in the conventional notation used during the preceding years. Special symbols that were utilized earlier to indicate logic representations included one that had the appearance of a thin-lined capital U, whereas another was the same symbol in inverted form. These and other such symbols have been replaced by the more conventional arithmetic signs, although in logic processes a $+$ or a $-$ sign does not have the same significance as it does in math equations.

In the delineation of logic expressions in digital electronics only a few symbols are needed because only a few basic circuits are utilized. In the complete system these few logic circuits are used numerous times in numerous combinations to perform switching and gating functions in calculators, computers, telephone networks, and so on. Logic system terminology and methodology will appear strange to one unfamiliar with modern practices and symbols. One does not need to know electronic circuitry to design an overall logic system because illustrative representations differ from its discrete electronic circuit counterpart. For instance, a common logic circuit is one in which an input designated as A will be equally processed as one designated as B or as C, and so on. Thus, the key logic word here is *or*, since the logic expression is A or B or C, and these, entered singly or simultaneously, will appear at the output.

One of the logic symbols for the OR function is shown in Fig. 6-4(a). The input lines are at the left and the output lines are at the right in conventional schematic sequence. More than two inputs could be utilized and a typical three-input representation is shown in (b). Note the output expressions are separated by the mathematical plus (+) symbol. In logic designations the plus sign used in this fashion is designated as the *logic connective* representative of the logic OR function. It does not indicate a math additive process, although a combining function does prevail. If, for instance, the literal A represents binary 1010 and B represents 101, the output binary number would be 1111. When numbers such as 101 and 110 are entered, however, there is no add function, only a combining process whereby 1s that occur simultaneously simply produce a 1 output. Thus, 101 and 110 would produce 111 at the output, instead of 1011. When true sums are needed, a special symbol such as

Sec. 6-6 Some Logic System Aspects

shown in (d) is used to represent an arithmetic circuit that performs the function of $1 + 1 = 0$. This symbol represents an *exclusive-or* function.

Another logic process is based on a coincidence factor, that is, a signal appears at the output *only* if coinciding signals are applied to the input. Such a logic circuit is sometimes called a *coincidence gate* or an AND circuit. As with the OR circuit it can have a number of inputs. A two-input AND circuit is shown in Fig. 6-5(a), while a four-input unit is shown in (b). Note that the plus symbol is no longer used. Instead, the multiplication function is indicated such as denoting multiplication by placing letters close to each other as in ABC or by utilizing a raised dot between them such as $A \cdot B \cdot C$. The multiplication sign does not indicate a mathematical process but rather the logic involved in an expression such as A and B and C, and so on. The logic decreed here is that an output is obtained *only* if an input signal appears in coincidence at each input terminal. Typically, this is exemplified in (c) where the A input is 1011 and the B input is 1001. Since coincidence prevails only for the rightmost and left-most digits, the output expression is 1001. The AND function is useful for gating specific binary bits into or out of circuits as required. For instance, in (d) a train of 1s appears at the upper input, but only a single 1 appears in fourth place for the lower input. Consequently, coincidence only appears for the fourth place digits producing an output as shown.

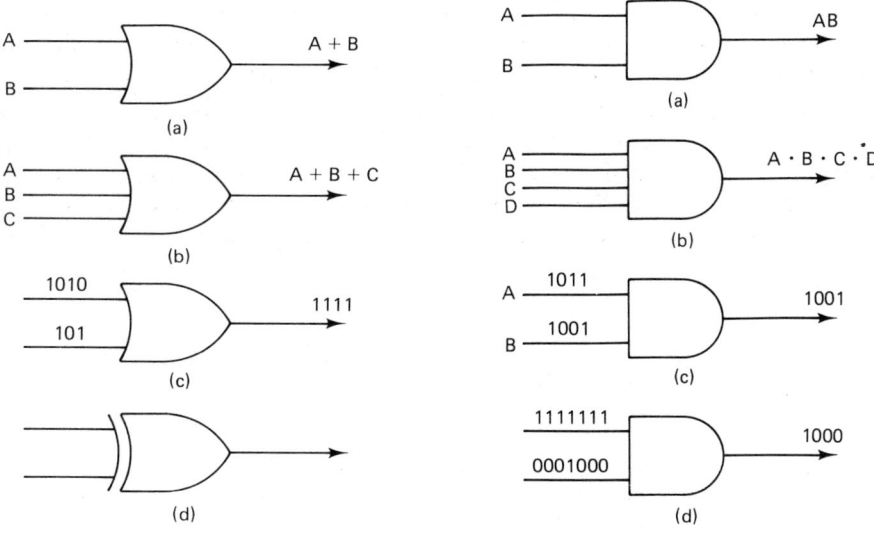

Figure 6-4   Figure 6-5

Earlier in Sec. 6-2 the complementary factor was introduced. The complementary process is essentially a logic negation. Thus, if we have an expression such as A+B+C and wish to indicate that each letter symbol is negated, we would place an overbar over each and the final expression would thus be $\overline{A}+\overline{B}+\overline{C}$. If the entire expression is negated it becomes $\overline{A+B+C}$, and since we have now negated the logical connectives, the expression is actually $\overline{A}\cdot\overline{B}\cdot\overline{C}$. To indicate the negating process, the OR-circuit symbol is altered by placing a small circle at the output as shown in Fig. 6-6(a). When the OR circuit is negated in this fashion it is termed a NOR circuit, indicating the NOR function or *negated OR* function. A similar negation also involves an AND circuit as shown in (b), thus forming a NAND circuit. Various combinations of the logic gates are utilized and a typical example is shown in Fig. 6-6(c), where the output from an OR

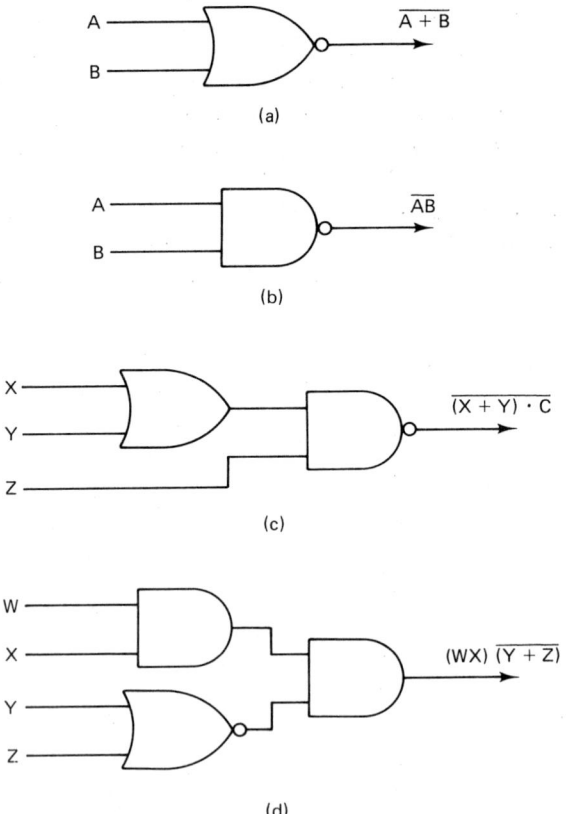

Figure 6-6

Sec. 6-7   A Taste of Boolean Algebra   89

circuit is applied to one input of a NAND circuit, and an independent input to the NAND circuit accepts the input designated as Z. Thus, the logic expression is X+Y·C, as shown. Another example is shown in (d) involving three logic gates. Inputs W and X are applied to an AND circuit but Y and Z are applied to a NOR gate. The outputs from these two are applied to another AND circuit to produce the logic output expression of (WX)($\overline{Y}$+$\overline{Z}$).

6-7. A TASTE OF BOOLEAN ALGEBRA

Earlier, in Sec. 6-6 some introductory aspects of Boolean symbolic algebra were presented. Present-day logic systems employ the same basic rules and laws formulated in the early days of the inception of this symbolic algebra. The 0 and 1 in binary arithmetic are called identity elements because when they are used in combination with a letter symbol it leaves the latter unchanged. Thus, this precept forms two laws of identity.

    1. The addition of 0 to a literal leaves the latter unchanged as A+0=A.

    2. The multiplication of a literal by 1 leaves the literal unchanged as A·1=A.

The laws of complementation state: To complement an expression change all 1s to 0s and all 0s to 1s. Change each logical connective that was negated from AND to OR, or from OR to AND. Change a letter symbol such as A to $\overline{A}$ or $\overline{A}$ to A. Other similar laws have been formulated. Typical is the one termed De Morgan's theorem, which states that the complement of a product of literals is equivalent to the sum of the separate literal complements. However, the complementation of the sum of literals is equivalent to the product of the separate literal complements. Thus, $\overline{A \cdot B}$ = $\overline{A}$ +$\overline{B}$, while $\overline{A+B}$ = $\overline{A} \cdot \overline{B}$.

From the foregoing it is evident that Boolean algebra appears complex primarily because of the difference between it and standard numerical algebra. In the latter, for instance, an expression such as $A \cdot A$ represents A^2, whereas in Boolean algebra no exponents are used and the coefficients are only 1 and 0. Hence AA as well as the expression $A+A$ becomes only an A. Thus, in Boolean algebra, circuitry can be simplified since the logic A+A is a redundant expression and can be simplified to a single A thus eliminating a switch or gate circuit. Consequently, many complex and numerous circuits can be eliminated or reduced by using Boolean algebra. As another example, $A(A+B+C+D)$ can be simplified by representing it by a single literal A because if A alone is valid, A in combination with other literals is superfluous. Similarly, an expression such as $A(A+B+C) \cdot X + (XYZ)$ in simplified form reduces to AX.

If an input designated as A is combined with the numeral 0 in an AND gate we do not obtain coincidence and hence $A \cdot 0 = 0$. Also $A \cdot \overline{A} = 0$. A few other representative logic expressions are:

$$A+(AB) = A$$
$$A+(\overline{A}B) = A+B$$
$$A(A+B) = A$$
$$(A\overline{B})+(AB) = A$$
$$(A+B)(A+\overline{B}) = A$$
$$A(\overline{A}+B) = AB$$
$$A+\overline{A} = A$$
$$A+0 = A$$

6-8. VENN DIAGRAMS ARE USEFUL

A useful convenience is to indicate the logic of an expression in diagram form. Two basic types are utilized, one in block formation with internal squares and the other using a block with circle representations as shown in Fig. 6-7. Such a representation is termed a *Venn diagram*. Either method can be used to represent a variety of logic expressions. The block with squares utilizes horizontal rows to repre-

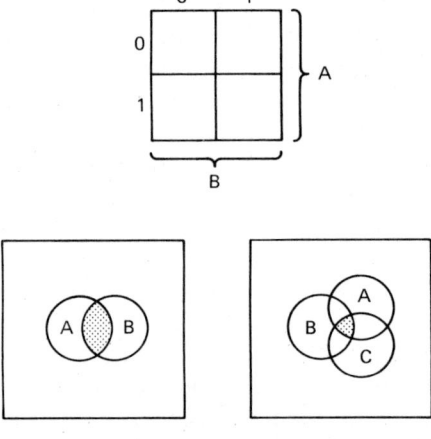

Figure 6-7

sent logic 1 or logic 0 for the literal A. Vertical rows represent the 1 and 0 for the literal B. As also shown in Fig. 6-7, the Venn diagrams utilize circles to represent two or three letters.

For both representations, shaded areas are used to indicate specific logic conditions. As shown in Fig. 6-8, the lower squares are shaded to represent A, whereas the right-hand vertical rows are shaded to represent B. If logic-0 is to be represented (a negated A or B), the shading is the inverse of what it was for A and B, as shown. Thus, the upper horizontal squares are shaded to represent $\overline{A}$, whereas the left vertical squares are shaded to represent $\overline{B}$. Combinations of A and B can be depicted by shading specific squares. As shown in Fig. 6-9, the representation for AB is achieved by shading the lower right square to represent the logic function A AND B. For the logic A+B three squares are shaded, leaving the upper left square unshaded. For the logic expression $\overline{A+B}$, which is also $\overline{A} \cdot \overline{B}$, only the upper left square is shaded as shown. For the expression $\overline{AB}$, which is also $\overline{A}+\overline{B}$, all squares are shaded except the lower right.

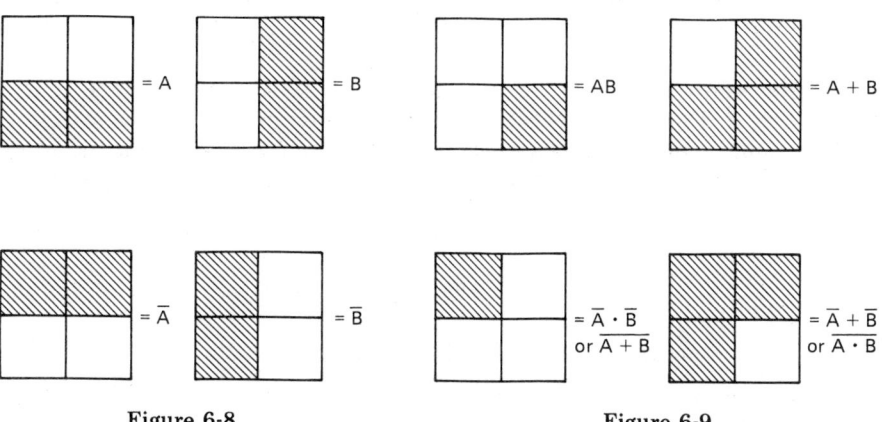

Figure 6-8   Figure 6-9

The Venn diagrams are named after John Venn, the nineteenth-century English logician who utilized the basic concepts originated by Leonhard Euler, the famed eighteenth-century Swiss mathematician. Venn solidified the concept of the intersection of two sets to demonstrate the logic expression. Venn thus advanced symbolic logic by providing a simple diagrammatic method of identifying logic expressions. Typical examples of the Venn diagrams are shown in Fig. 6-10. Note that for the expression A+B two intersecting shaded circles are shown. For the *and-gate* logic expression of AB two unshaded circles are shown, but the area of their intersection is

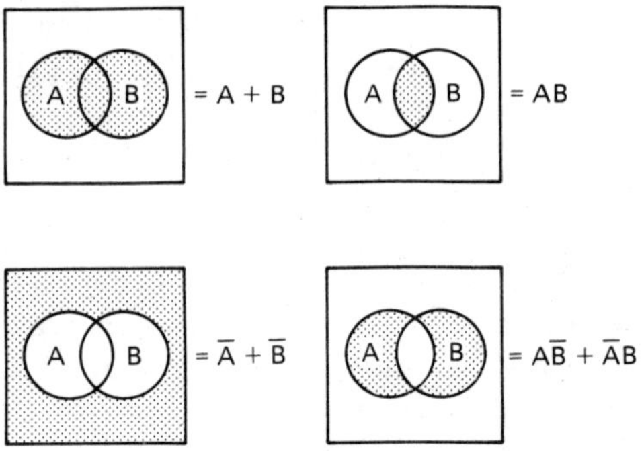

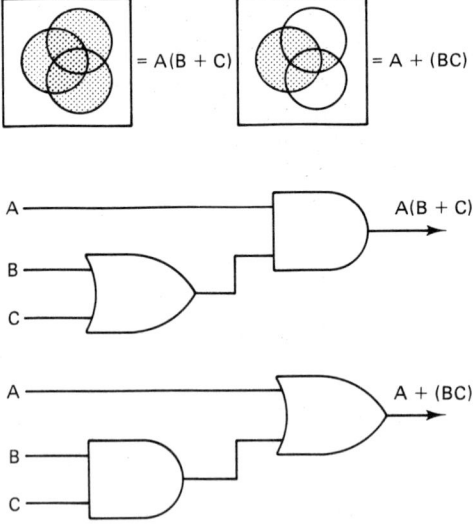

Figure 6-10

shaded. For the expression $\overline{A}+\overline{B}$ we show two unshaded circles against a shaded block background. The same diagram also applies for the equivalent logic expression of $\overline{A \cdot B}$. For an expression such as $A\overline{B}+\overline{A}B$, two shaded circles are shown, but their intersection is unshaded. For the same $A\overline{B}+\overline{A}B$ expression the block representation with squares would have the upper right and the lower left squares shaded.

The Venn diagram for the expression $A(B+C)$ is shown at the upper left of Fig. 6-11. Note that three shaded circles are overlapping with heavier shading for overlapping segments. The logic symbol representation for this expression is also shown in Fig. 6-11. For the

Figure 6-11

Sec. 6-8   Venn Diagrams Are Useful

expression A+(BC), we again have three intersecting circles with only one circle shaded, but the intersection of circles D and C also is shaded. The representative logic diagram utilizing AND and OR gating is shown also.

The foregoing represents a brief discourse on the utilization of diagrammatic representations for logic expressions. It is included herein to illustrate the oddities and offbeat aspects of representations utilized in computer design. All the subjects touched on in this chapter involve much more complex configurations and representation in actual practice and would require several volumes to cover all subjects in depth. More lengthy logic expressions are involved plus multisquare logic mapping practices. All procedures, however, are based on the peculiar aspects of binary and symbolic logic that have been discussed briefly in this chapter.

7
BASICS OF COMPUTER LANGUAGES AND PROGRAM STRUCTURES

INTRODUCTION

One of the great advantages of a digital computer is its versatility in accepting and processing program variations. A variety of program types have been devised over the years, and a number of the earlier types were designed specifically for a certain area such as data processing in business or equation processing in scientific calculations. Over the years, however, program types were not only simplified but were made more versatile in programming diversification. Given sufficient storage capabilities and a flexible program type, there is virtually no limit to the type of instructions that can be given the computer nor the complexity of the problem assigned to it.

Mathematical problems that would require hours of work by a trained mathematician can be performed in minutes by modern computers because of the incredible speed with which the computer handles a particular problem. In computerized chess games, for instance, the computer can search through millions of potential moves and come up with an appropriate move well within specific time limits set by standard chess rules. Similarly, the computer can be programmed to perform all the intricacies of a simulated space adventure-type game complete with a visual display in color, moving objects, and sound effects. Again, millions of functions are performed in a few seconds, and at the same time the computer maintains a constant running score of the status of the game being played.

In this chapter some programming factors and examples are given to illustrate some of the lighter aspects of programming. The purpose here is not to teach the specifics of FORTRAN, BASIC, or other program language types. The manual of the computer used will cover the details of rules, methodology, and applications of a particular program type suitable for the computer in use. Rather, the structural foundation of a given program is covered in this chapter. Once you are familiar with flow-charting principles, the diagram drawn can be used with any program language type.

Initially, an overview of some basic program types is given for a clearer understanding of computer programming concepts. Next the aspects of structured programming are covered in detail to indicate how a problem can be illustrated in a sequential symbol structure which can be utilized, when finalized, for any program type. In later sections specific examples of the planning and construction for puzzle and game programs are discussed and illustrated.

7-1. SOME FUNDAMENTAL PROGRAM TYPES

The instruction and data regarding specific processes to be performed are termed a program. A computer program must spell out in detail the exact sequence of steps the computer must take. In addition the computer must be given the data necessary to perform the steps required. As mentioned in the previous chapter in Sec. 6-4, the particular computer language utilized is fed to the computer via a translating device that is programmed to convert the high-level computer language into the low-level machine language (binary) utilized within the computer. Such translation devices are either assemblers or compilers, and in most instances they are designed so they accept only a specific program language.

There are a number of program languages available for modern computers. One of the earliest program types is FORTRAN, and this word is an acronym devised by utilizing portions of the phrase *formula translator*. The FORTRAN computer language permits the statement of problems by using an equation structure very near that utilized in basic mathematics. As with all programs, specific rules must be observed in setting up a FORTRAN program. Each statement permitted must be used in a specific manner as must also be the particular words within a statement. Since the letter X can be mistaken for an algebraic expression as well as a multiplication sign, an asterisk (*) is utilized in FORTRAN to depict multiplication. For exponentiation a double asterisk (**) is used. Division uses the solidus (/), whereas addition and subtraction are simply the symbols + and −.

Variables must be expressed in capital letters and an equation is written on a single line. Parentheses must be used if more than one operational symbol appears between two variables. In the use of parentheses it is necessary to have an equal number of right and left ones for any single line equation. FORTRAN, as with other programs, must be written in strict prescribed form or the compiler translator will be unable to process the statements. Letter symbols are provided for such arithmetic operations as cosine (COS), sine (SIN), square root (SQRT), and so on. As with all high-level programs numerous other statements are utilized, including instructions for decision modes, print out, read in, and so on.

Another early program type is that referred to as COBOL,

which is an acronym formed by utilizing initial letters of the phrase *common business-oriented language*. This program language utilizes basic English in statements, as well as letters, numbers, and punctuation marks. As with all program languages specific rules must be observed for program formation. Once written, the program can be understood readily, but the untrained person cannot write the COBOL program unless he/she learns the specific rules that apply. For instance, a valid statement would be COMPUTE ANNUAL IN SALARIES = 12* MONTHLY PAY. Substituting the word *calculate* or *execute* would invalidate the statement. The asterisk is for multiplication as in FORTRAN. The COBOL computer program language has four fundamental divisions: *identification, data division, environment,* and *procedure.* The initial three divisions identify the particular program plus the nature and source of the data. The last division spells out the sequential steps required to process the data.

Another high-level language is that identified by the symbol PL/1 (program language 1). It combines some of the fundamental aspects of COBOL and FORTRAN and was one of the first program types designed for universal usage and versatility. It has a somewhat simpler program structure than COBOL or FORTRAN. Common words are used for execution including the word PUT for placing data into storage and GET for retrieving data from storage. The asterisk and other signs used in FORTRAN and COBOL are common to PL/1.

Another program closely allied to FORTRAN is ALGOL. The designation is an acronym derived from the phrase *algebraic-oriented language*. It was intended initially to serve as a simplified version of FORTRAN with increased versatility. In subsequent versions of FORTRAN, however, the advantages held by ALGOL became minimal, particularly with FORTRAN IV.

Another special language was APL, derived from the phrase *a programming language*. The approximately sixty primary programming functions of APL are divided into two sections: scalar and mixed. The language uses Greek letters, some special symbols for operations and functions, and alphanumeric designations.

7-2. MISCELLANEOUS PROGRAM LANGUAGES

There are a number of other computer program languages in addition to the types discussed in Sec. 7-1. Many of the newer types were originally designed for early minicomputers to overcome the handicaps of limited storage facilities. As the state of the art improved, vastly increased storage capabilities were realized, thus permitting usage of the well-known languages such as FORTRAN, COBOL, and BASIC (the latter discussed later). One of the various languages developed for later computers is SMAL, which is derived from the phrase *structured macro-assembly language*. The MACRO portion of the phrase defines a group of assembly-language mnemonics (aid to memory). This computer language simplifies the structured program principles discussed more fully in Sec. 7-4.

Several computer programming languages have special names, such as PASCAL and ADA. Instead of being acronyms, these designations are *eponyms*, because they are named after people. The PASCAL computer language is named after the famed French mathematician, Blaise Pascal (1623-1662). The PASCAL language is highly structured and has features that permit the preparation of sophisticated computer programs. Complex data structures can be formed much more easily than with previous program languages. The language is essentially one that permits programs to be formed in two sections. One section identifies the program as well as specifying the variables. The program body (termed a block) follows, and this is sectionalized into six additional portions. Of these six, the first four declare the labels of the program, the constants utilized, the data types, and the variables to be used. The fifth portion identifies an actual function, and the last section contains the code that initiates the specified procedures. The PASCAL language has a few advantages over FORTRAN and BASIC and often results in the reduction of programming time. As with most of the programming languages, improvements over the original version have been formulated.

The ADA language mentioned above is an eponym for Augusta Ada Byron, who apparently was the first person to program a computer. She was an assistant to the famed Charles Babbage (1792-1871) on his mechanical device that was the first true computer unit capable of performing basic mathematical processes and problem solving. The ADA language also incorporates a structured design as with PASCAL. The ADA language can be readily utilized for constructing specific programs without problems associated with the incompatibility that might otherwise prevail because of the expanded program base.

7-3. BASIC COMPUTER LANGUAGE

One of the most popular program types that followed the earlier programs of FORTRAN, COBOL, and PL/1 is the program language named BASIC, which is an acronym derived from the phrase *begin-*

ner's all-purpose symbolic instruction code. This is a versatile language, and although it uses many of the principles of FORTRAN, it can be used for business-type data processing as well as for solving problems in general mathematics, engineering, and science. It has been widely used not only for home computers but also for the more sophisticated business-type machines.

As with other programming languages, there have been periodic alterations made for the fundamental BASIC format. Whereas earlier versions utilized the word "LET," for instance in a statement such as LET X = ∅, in modern versions the entry is simply X = ∅. Where some earlier versions use REMARK for entering reference data, modern versions use the abbreviated REM. Similarly, some computer models use an upper pointing arrow (↑) for exponentiation as in the original (Dartmouth College) BASIC, while others utilize the ∧ sign. Similarly, some computers permit the statement PRT instead of PRINT or INP instead of INPUT. Except for such minor points, however, the general format of the BASIC language has the same fundamental aspects as the original.

7-4. FLOW-CHART STRUCTURES

When programs are written for a specific computer it usually means that only one program language is acceptable. Thus, if a number of programs already written with FORTRAN were also to be used on another computer limited to the BASIC computer language, it would require someone familiar with FORTRAN notations as well as BASIC in order to rewrite the original FORTRAN program into the BASIC programming language. Thus, the format of a complex data process or mathematical problem (also called the algorithm) would have to be reworked for another programming language. Hence, to reduce reprogramming time and to simplify the adoption and rewriting of existing programs into another program language, *structured programming techniques* have become common. With the latter, the algorithm or program sequence is structured along logical flow paths. With this method symbols are used to signify certain conditions, and these symbols are drawn in a sequential order following the algorithm process. Once such a flow chart has been structured, it can be used to write an appropriate program in any programming language without having to decipher an existing program.

Flow-charting techniques have been utilized for some time in the computer sciences, although the more formal concept of structured coding was developed later and is a more rigid process. For the structured programming, two basic guidelines are utilized. One such guideline specifies that each programming module should have only a single entry and a single exit point as shown in Fig. 7-1. Here squares or rectangles are utilized and the sequence progression is from left to right. The second guideline specifies that the sequential flow of the control structures in the program or in a diagram module be limited to only three specific structures. (This limit was established because the three structures have been found to be sufficient for expressing any algorithm.)

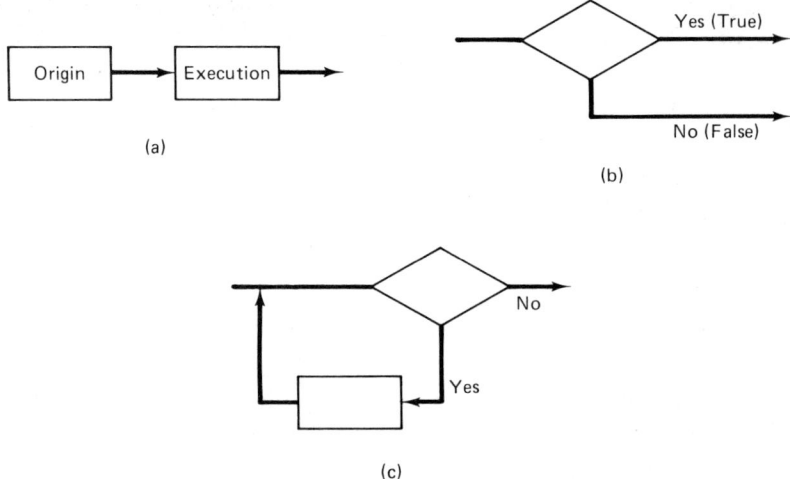

Figure 7-1

A decision structure is identified by a triangle as shown in Fig. 7-1(b). Here we have a single entry and alternative outputs. The output structures consist of a *yes* output also termed *true*, and a *no* output also termed *false*. Since only one of these is selected by the computer, we still have only a single entry and a single exit. This decision structure is of the type utilized for such program commands as IF, THEN, and ELSE-type commands. Loop structures can also be formed as shown in Fig. 7-1(c). Here if a *no* situation prevails the output is progressive. If, however, a *yes* position prevails, a loop function is undertaken as shown.

Structured program logic diagramming is reducible because a single block replacement can be made, for instance, with a loop function as shown in Fig. 7-2. Here is a decision function and a loop plus the entry block that can all be reduced by a single block as shown. Some variations in diagram usage may be found for different

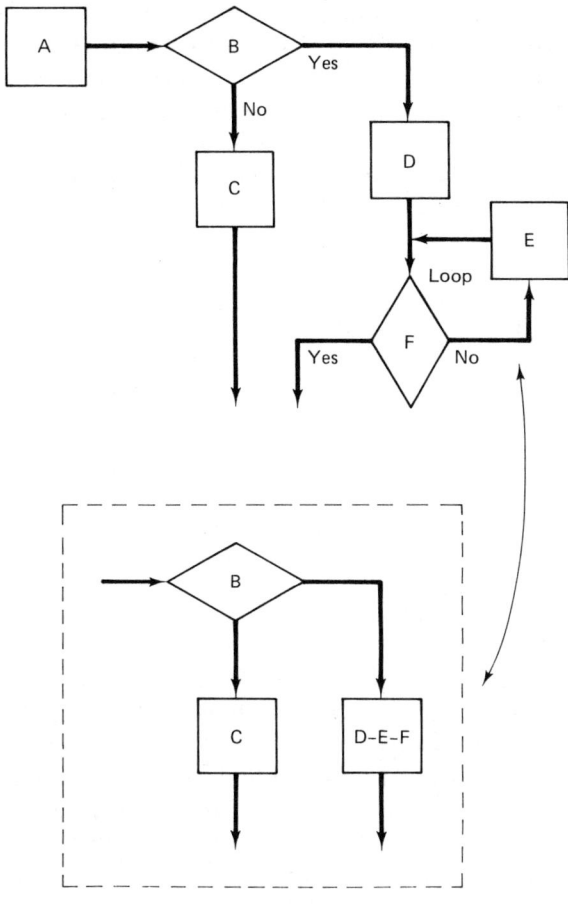

Figure 7-2

programmers. A circle may be used for data origin or as a *start* or *stop* symbol. Some may use rectangles instead of squares. The basic sequential flow, however, is still the same.

7-5. PROGRAM CHARTING AND EXAMPLES

The flow-chart blocks are identified by writing within them the operation to be performed. Sufficient data are included to serve as guidelines for making up a specific program. In the simple programs there may not be any decision processes, hence the flow chart and program are easily constructed. As an example, assume you wish the computer to convert temperature readings from Fahrenheit to both Celsius and Kelvin. To do this we must refer to the conversion formulas. Fahrenheit measurements are converted to Centigrade by the equation $C = 5/9\ (F - 32)$. Kelvin temperatures are converted from Celsius by the equation $K = C + 273$. For this program we wish to enter a Fahrenheit temperature in degrees and want a readout of the equivalent Celsius and Kelvin values.

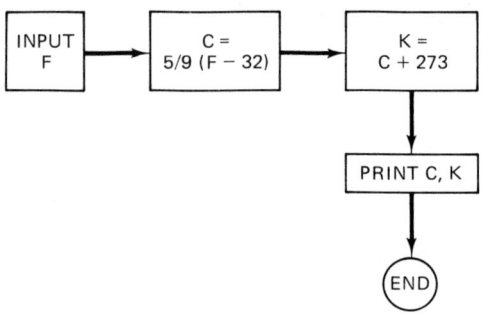

Figure 7-3

Utilizing the above formulas permits us to construct a flow chart as shown in Fig. 7-3. Here we have an input X, which constitutes the degrees of temperature to be converted from the Fahrenheit scale to the others. The next block sets X to equal F, followed by the equation for Celsius and that for Kelvin. Now the computer has all the data, and it is asked to print the values of the original F entry as well as the C and R. From such a flow chart a program can be written for any computer language. For the BASIC computer language the program would appear as shown in Program 7-1.

```
 95 REM: TEMPERATURE CONVERSION
100 PRINT "DEGREES FAHRENHEIT IS"
    :INPUT F
105 PRINT
110 C = 5/9*(F-32)
115 K = C+273
120 PRINT "DEGREES K IS" :K
125 PRINT "DEGREES C IS" :C
130 END
```

Program 7-1

Thus, if we entered a Fahrenheit of 32 the computer would print:

Fahrenheit
32
Degrees K is 273
Degrees C is   0

Similarly, if F = 75 the computer prints:

Fahrenheit
75
Degrees K is 296.888
Degrees C is 23.888

And finally, if F = 212 the printout is:

Fahrenheit
212
Degrees K is 373
Degrees C is 100

    Most programs will be much more extensive than the preceding and hence will require a greater number of identifying blocks. If, for instance, we needed a flow chart that would have the computer undertake the complete operations for a magazine subscription process, it would appear as shown in Fig. 7-4. Here, the first box at the upper left indicates the computer is to read from storage the sequential listings of magazine subscriptions. Thus the computer samples the first subscription and then checks whether or not the date indicates the subscription has expired. If it has not expired, the computer prints the magazine address label for mailing purposes and then automatically checks whether any listings remain. If the list is not exhausted, the computer selects the next subscription listing and again checks the date to see whether or not it has expired. If the subscription has expired, the computer is to print an expiration notice (which will be sent to the subscriber) and an address label for mailing the notice. The next command is to remove the expired name from the list, although some companies may hold a subscription for a month or two after the expiration date. The process continues until the list is exhausted, at which time the program command is to stop the operational process.

    Mathematical equations would have similar decision segments usually representing an IF command in programming. Such commands represent a branch operation. Thus, if the value is not a specific one originally designated, the computer is to continue suc-

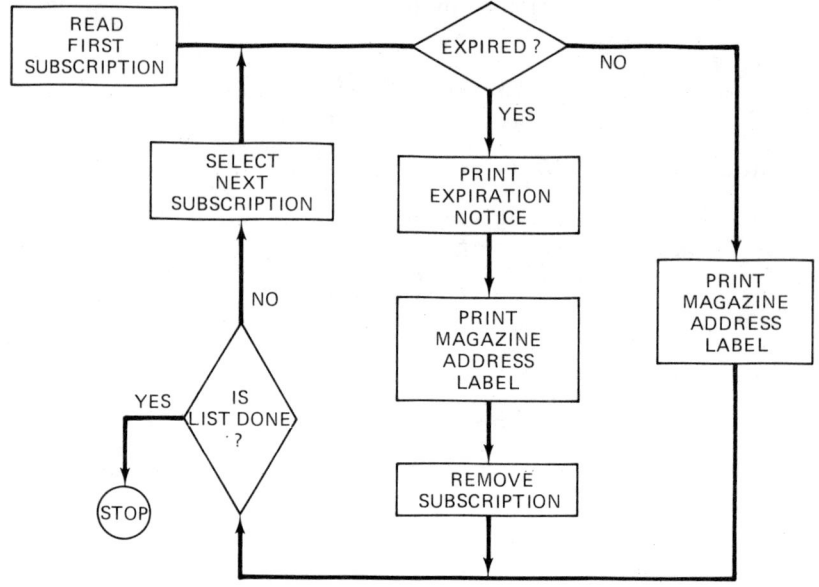

Figure 7-4

cessive operations. If the value is a specific one, a branch command is initiated to some subsequent (or previous) program step. Note the branch commands incorporated in the BASIC language statements shown in Program 7-2.

```
100 REM: NUMBERS TO 64
110 X=1
120 IF X > 64 THEN 160
130 PRINT X
140 X=X+1
150 GO TO 120
160 PRINT "LIST COMPLETE"
```

Program 7-2

For Program 7-2 the computer is asked to print numbers from 1 to 64. Thus X is initially set to 1 at memory designation 110. Next, at 120, if the number has not reached the designated limit, it prints out the X at 130. Then X is set to increase by one and at 150 a branch command sends the operation back to 120. Finally, when X is greater than 64 the branch is to 160, where the printout says "LIST COMPLETE" and the program run stops.

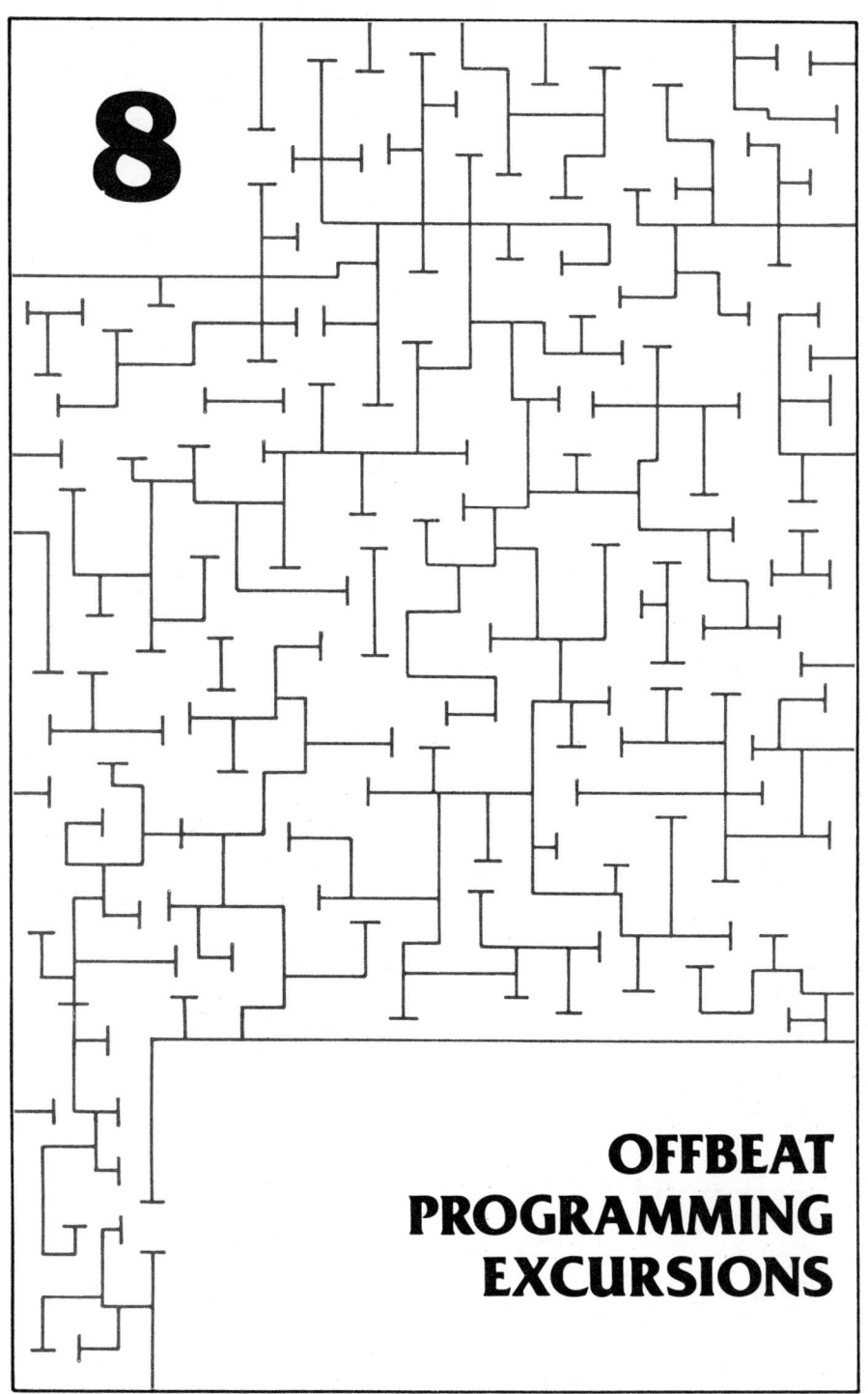

8
OFFBEAT PROGRAMMING EXCURSIONS

INTRODUCTION

The popularity of home computers can be attributed not only to their computing and data storage capabilities but also to their ability to accept preprogrammed cartridges or tapes for game-playing purposes. With any home computer, however, there are many easily constructed programs that can be entered for entertainment purposes. Such programs are also somewhat different in structure from the representative video-game format. Some of these entertainment-type programs are sampled in this chapter, including several structured around some of the tricks covered earlier in Chapter 3.

Each program in this chapter is accompanied by a description of the basic format and some of the special techniques used in formulating it. Except for an example of the FORTRAN program in Sec. 8-1, the programs are written in the BASIC computer language. As discussed in Sec. 7-3, some slight differences are encountered in the utilization of the BASIC language by different computer manufacturers. Consequently, some changes in statements may be necessary to run the programs on a particular computer. Primary differences may be present in random-number commands, certain loop progressions, and so on. The step-by-step sequence, however, will be essentially as presented in this chapter. Each program shown has been run successfully on a computer with the consequent printout as represented herein.

8-1. A BIGGER PIECE OF PI

Mathematical pi is extensively used in a number of equations, particularly of the scientific nature. Generally the much abbreviated form of 3.14 is utilized. In some instances students have divided 22 by 7 to obtain an extension of the numerals beyond the decimal point. With the advent of calculators, however, a more accurate representation was easily obtained and a typical eight-digit readout is 3.1415926. Certainly, it would be a rare occasion where a mathematician needs a larger number, so all the foregoing procedures suffice when pi is required. It has always intrigued me, however, regarding the methodology utilized for ascertaining an infinite number of places beyond the decimal point. Of the several procedures used,

I've always thought the following method was particularly interesting for solving pi to as many numbers as desired:

$$\text{pi} = 4(1 - 1/3 + 1/5 - 1/7 + 1/9 - 1/11 + 1/13 - 1/15 \ldots n)$$

Note the curious aspect of this formula. There is a sequence of subtractions and additions involving odd-number fractions. At the time I ran across this equation I had access to a computer using FORTRAN and wrote a program for this sequence. The complete sequence is shown in Program 8-1. Initially I set the program down so that $A = 3$ and $B = 5$. Next the program is given a portion of the equation that states that $Y = 1 - 1/A + 1/B$. I then instructed the computer to add 4 to A and B, successively, so that I would have a rising amplitude of odd numbers. The computer is then instructed to form the equation a specific number of times and then to multiply the results by 4. Next the computer is told to print the answer so obtained in the format specified, printing thirty characters of the final answer.

```
50 PROGRAM PYE (INPUT,OUTPUT)
55 CALL TYPEIT
60 A = 3.
65 B = 5.
70 Y = 1 - 1/A + 1/B
75 DO 9 J = 1,1000,1
80 A = A + 4.
85 B = B + 4.
90 9 Y = Y-1/A+1/B
95 PI = 4. * Y
100 PRINT 10,PI
105 FORMAT (2X,F30.28)110 END
119 END
```

Program 8-1

Although the computer printout extends for twenty-eight digits beyond the decimal point, the mathematical expression is not an accurate pi. Next I introduced the same program into the computer but changed line 75 to:

75   DO 9 J = 1, 100000, 1

Thus, steps 80, 85, and 90 were taken 100,000 times instead of only 1000 times as in the original program. The new printout was:

3.1415976520945889660652028397

Note that although we have maintained the same number of digits in the readout we now have a much more accurate presentation of pi to twenty-eight places beyond the decimal point. However, the inaccuracy is still present because the numeral in the sixth place after the decimal point should be a 2. For twenty-eight places be-

Sec. 8-2   Computer Has a Trick   111

yond the decimal point we should obtain the following:

$$3.141592653589793238466 8394798$$

The reader might be interested in carrying the equation extension to some 200,000 or 1,000,000 steps for lines 80, 85, and 90 to ascertain the degree of accuracy that might be achieved.

There are several valid methods for calculating pi and another transcendental version is illustrated in line 20 of Program 8-2 using BASIC. Note that initial values are programmed for X and pi in lines 10 and 15. In step 30 we progressively implement the X. The last statement (line 30) forms a loop to line 20 (the body of the equation) for repetition. When this program is run you will see a continuous changing display as the calculation proceeds for as long as you want. It can be stopped any time by pressing the *run stop* key on the computer to find the value of pi at that time. The longer the program is run the greater the accuracy of pi, as was the case with Program 8-1.

```
 5 PRINT "PI"
10 X = 1
15 PI = 0
20 PI = PI + 4/X - 4/(X + 2)
25 PRINT"PI"
30 X = X + 4
35 GO TO 20
```

Program 8-2

8-2. COMPUTER HAS A TRICK

Mathematical tricks are easily programmed because they consist of a series of instructions for manipulating numbers in a specific sequence. When the sequence ends, a simple equation produces the original number thought of by the participant. A typical example of this type is Program 8-3. The latter is based on the trick entitled "This One's Easy" discussed in Sec. 3-10. In referring back to that section you will note that the participant selects a number secretly and is told to multiply it by 3 and then by 2. The result is then divided by 12 and finally 5 is added to the quotient. When the answer is disclosed you multiply it by 2 and subtract 10 to find the original number. These procedures are set down as shown in Program 8-3, and the program sequence simply consists of instructing the participant in the mathematical steps previously outlined. When the participant reaches the final number, it is typed into the computer and the latter displays the original number.

The graphic symbols used in lines 20, 35, 55, etc., can be

```
10 GOSUB 160
15 PRINT "/\/\/\/\/\/\/\/\/\         /\/\/\/\/\/\/\/\/\"
20 PRINT "■ COMPUTER HAS A TRICK "
25 PRINT "/\/\/\/\/\/\/\/\/\         /\/\/\/\/\/\/\/\/\"
30 GOSUB 125
35 PRINT "■USING A CALCULATOR    PUNCH IN A SELECTED    SECRET NUMBER"
40 GOSUB 125
45 PRINT "MULTIPLY YOUR SECRET    NUMBER BY 3"
50 GOSUB 125
55 PRINT "■    MULTIPLY AGAIN    BUT NOW BY 2."
60 GOSUB 125
65 PRINT "■DIVIDE PRODUCT BY ■12"
70 GOSUB 125
75 PRINT "■ADD 5 TO YOUR ANSWER"
80 GOSUB 125
85 PRINT "■NOW TYPE THE NUMBER    APPEARING ON THE       CALCULATOR INTO COMPU-TOP
"
90 GOSUB 160
95 PRINT "THE COMPUTER WILL    DISPLAY YOUR ORIGINAL SECRET NUMBER": INPUT N
100 Y = (N*2)-10
105 GOSUB 160
110 PRINT "■YOUR ORIGINAL NUMBER    WAS " Y
115 PRINT
120 PRINT "■END OF TRICK"
122 END
125 PRINT
130 PRINT
135 PRINT
140 PRINT
145 PRINT
150 PRINT "■PRESS RETURN KEY    TO GO ON": INPUT X
155 RETURN
160 PRINT
165 PRINT
170 PRINT
175 PRINT
180 RETURN
```

Program 8-3

omitted if desired. These symbols (on the Commodore V-20 computer) initiate color displays for the statements involved. Two subroutines are used for spacing purposes as shown. In line 95 an input N is requested, which is utilized in line 100 to find the original number. The latter is now displayed by the print command in line 110.

Computers are designed to accept programs wherein several instructional statements can be included in a single-number entry. Some computers require that such instructions be separated by a colon and others require a solidus (/). Spaces can also be eliminated and other shortcut procedures utilized to form a concise program with a minimum of storage space requirements. Such bunching together of program instructions is sometimes termed *crunching* and is widely used by professional programmers.

For the purpose of presenting steps that are readily understood, Program 8-3 and the others in succeeding sections are not crunched. Thus, successive PRINT commands are given for spacing purposes, although TAB and SPC (space) commands are usually available. Utilizing the crunch principle the start of Program 8-3 could appear as:

Sec. 8-3   The Thirteen Trick Program

```
20  PRINTSPC(89)"COMPUTER HAS A TRICK":GOSUB 125
25  PRINT"USING A CALCULATOR PUNCH IN A SELECTED SECRET
    NUMBER:GOSUB 125
30  PRINT "MULTIPLY YOUR SECRET NUMBER BY 3":GOSUB 125:PRINT
    "MULTIPLY AGAIN BUT NOW BY 2"
50  GOSUB 125
    etc.
```

Now the subroutine 125 can be reduced from the print commands shown from 125 through 155 (the RETURN) to just two lines:

```
125 PRINT SPC(88)"PRESS RETURN KEY TO GO ON":INPUT X
130 RETURN
```

Because a crunched program listing is much more difficult to copy and enter via the computer keyboard, the programs in this chapter contain statements primarily in single instruction sequence. Since these programs are rather short, no problems should arise regarding adequate storage space for the program within the computer.

8-3. THE THIRTEEN TRICK PROGRAM

Program 8-4 is a computerized version of the trick game described in Sec. 3-11, where the game could not be won by your opponent if he went first. Thus, in the Thirteen Game program shown the computer requires your opponent to go first and hence the computer will never lose. Consequently, this is more of a trick than a game, and the uninitiated may try in vain to win a game from the computer. The program is rather lengthy, although some of the PRINT statements are blank to provide for spaces between the printout material. The program has safeguards to prevent an invalid entry such as a 0 or a number higher than 3. Thirteen paper clips, buttons, or other items are used and after the participant takes the number of pieces he selects, the number is entered into the computer. Next you remove the number of pieces indicated by the computer display. After three plays the computer indicates the game's end and announces the loser (always the participant).

Program 8-4 also serves as a guide for those interested in expanding the program so the computer alternates with the participant in starting first. In the latter case, of course, the computer can lose if the participant finds out the trick as detailed in Sec. 3-11. If your computer has graphic capabilities you can also write a program displaying thirteen objects or patterns. You will note that Line 90 states

```
5 PRINT "      THIRTEEN GAME "
15 PRINT
20 PRINT "        RULES    "
25 PRINT
30 PRINT "YOU PLAY THE COMPUTER. EACH TAKES A TURN TO PICK 1,2,OR 3 PIECES"
35 PRINT
40 PRINT "WHOEVER IS LEFT WITH THE LAST PIECE LOSES"
45 PRINT
50 PRINT "WHAT IS YOUR SELECTION? TYPE IT AND PRESS RETURN": INPUT X
55 Y = Y + 1
60 IF X = 1 THEN 85
65 IF X = 2 THEN 100
70 IF X = 3 THEN 115
75 IF X > 3 THEN 140
80 IF X = 0 THEN 150
85 PRINT "I SELECT 3"
90 IF Y = 3 THEN 130
95 GO TO 50
100 PRINT
101 PRINT "I SELECT 2 THE SAME AS YOU"
105 IF Y = 3 THEN 130
110 GO TO 50
115 PRINT
116 PRINT "I SELECT 1"
120 IF Y = 3 THEN 130
125 GO TO 50
126 PRINT
130 PRINT
131 PRINT
132 PRINT "THE GAME IS OVER.  TO PLAY ANOTHER PRESS THE RUN AND RETURN KEYS"
135 END
140 PRINT
141 PRINT "IT'S AGAINST THE RULES TO SELECT MORE THAN 3 PIECES."
145 GO TO 130
150 PRINT
151 PRINT "YOU CAN'T SKIP A TURN BY SELECTING A ZERO"
155 GO TO 130
```

Program 8-4

IF $Y = 3$ THEN 130, which senses the third selection by both participants and hence the end of the game. The loop is then to 130 where the printout mentions that the game is over. The game can be changed to 17 by changing the 3 in Line 85 to 4. Similarly, the game of 21 would require Line 90 to say IF $Y = 5$.

8-4. COMPUTER TRICK NO. 2

Another trick similar to that performed by Program 8-3 is one based on the first trick discussed in Sec. 3-14. In that trick the participant selects any number and (using a calculator) adds 222 to it. Next the resultant sum is multiplied by 3, and the original number is subtracted from the product. Next 666 is subtracted from the answer, and the end result is shown to you. Because the original number is found by dividing the final number by 2, the computer steps are minimized as shown in Program 8-5.

To provide for adequate spacing between the instructions displayed by the computer, PRINT commands are utilized. To avoid

Sec. 8-4   Computer Trick No. 2

the necessity for entering in groups of PRINT statements, the latter are placed into a subroutine and called upon as required as was done for Program 8-3. The RETURN statement in the subroutine automatically returns the program to the point of origin of the instruction. The subroutine starting at location 125 contains an instruction to press the RETURN key. The other subroutine, which simply provides spacing, starts at memory location 160.

You will note the PRINT command at location 150 has an INPUT instruction. Actually no input is asked for or entered, but this line provides for a program break that is sustained until the RETURN key is depressed. This provides as much time as required for processing the instructions into a calculator. As with Program 8-3, any number can be selected by the participant at the beginning of the trick as long as the number is within the amplitude of the calculator and the computer.

```
10 GOSUB 160
14 PRINT "————————————     ————————"
15 PRINT "■    COMPUTER'S TRICK       NO. 2 "
16 PRINT
17 PRINT "————————————     ————————"
20 GOSUB 125
25 PRINT "■USING A CALCULATOR    PUNCH IN ANY SECRET    NUMBER.■"
30 GOSUB 125
35 PRINT "■NOW ADD 222 TO YOUR    SELECTED NUMBER"
40 GOSUB 125
45 PRINT "■MULTIPLY SUM BY ■3 "
50 GOSUB 125
55 PRINT "■NOW SUBTRACT YOUR ORIGINAL NUMBER■"
60 GOSUB 125
65 PRINT "■SUBTRACT 666 FROM THE   REMAINING   NUMBER"
70 GOSUB 125
75 PRINT "■NOW TYPE INTO THE     COMPUTER THE NUMBER    SHOWING ON CALCULATOR■"
79 GOSUB 160
80 PRINT
85 PRINT "■COMPUTER WILL THEN    SHOW YOUR ORIGINAL    NUMBER": INPUT N
90 Y = N/2
95 GOSUB 160
100 PRINT "■YOUR ORIGINAL NUMBER    WAS " Y
105 PRINT
110 PRINT "END OF TRICK"
112 END
120 PRINT
125 PRINT
130 PRINT
135 PRINT
140 PRINT
145 PRINT
146 PRINT "■    ●●●●●●↑●●●●●●    "
147 PRINT
150 PRINT"■PRESS RETURN KEY TO    GO ON": INPUT X
152 PRINT
155 RETURN
160 PRINT
165 PRINT
170 PRINT
175 PRINT
180 RETURN
```

Program 8-5

8-5. THE SHELL GAME

Another interesting program involving numbers is Program 8-6. Although the title is "The Shell Game," it is not related to "The Numerical Shell Game" discussed in Sec. 3-7. Actually Program 8-6 follows the pattern of the actual shell game with the concept that the computer hides a kernel of corn under one of three shells. In essence this is a game in which you are pitted against the computer. You get three points for each correct answer, and the computer gets two points each time you miss. Whoever reaches 24 points first wins the game.

Because this program has more steps than the ones described earlier in this chapter it is more complex. In addition it contains

```
8 PRINT
25 PRINT "     THE SHELL GAME      "
30 PRINT
31 PRINT
32 PRINT
40 PRINT " THE COMPUTER HAS PUT  ONE KERNAL OF CORN    UNDER ONE SHELL"
45 PRINT
50 PRINT "█ IT'S YOU AGAINST THE   COMPUTER !"
55 PRINT
60 PRINT "█ YOU GET 3 POINTS FOR   EACH CORRECT GUESS--"
65 PRINT "█ THE COMPUTER GETS 2    POINT EACH TIME YOU    MISS."
70 PRINT
75 PRINT "█ WHOEVER GETS THE 1ST   24 POINTS WINS."
80 FOR T = 1 TO 8000:NEXT T
85 PRINT
90 PRINT
95 Y=0: Z=0
100 N=INT(RND(1)*3)+1
105 PRINT
107 PRINT "█ THE COMPUTER HAS NOW  PLACED THE KERNAL     UNDER A SHELL"
108 PRINT
110 PRINT"█WHICH SHELL DO YOU     THINK IT IS UNDER?- TYPE NO.1, 2, OR 3"
111 PRINT
112 PRINT "█PRESS RETURN KEY": INPUT X
113 PRINT
115 IF N = X THEN 150
120 Y =Y+2:IF Y=24 THEN 175
125 PRINT "█ YOU MISSED--COMPUTER NOW HAS█"Y" POINTS █    AND YOU HAVE█"Z" POINTS"
127 PRINT
128 PRINT"█THE COMPUTER HAD PUT   THE KERNAL IN SHELL"N
130 GO TO 100
140 PRINT
150 PRINT"█YOU GUESSED CORRECTLY-YOU WON 3 POINTS. "
152 PRINT
153 PRINT "█THE COMPUTER HAD PUT   THE KERNAL IN SHELL"N
155 Z = Z+3:IF Z=24 THEN 175
160 PRINT"█YOUR TOTAL NOW IS "Z"    AND THE COMPUTER'S    IS "Y
165 GO TO 100
175 PRINT "█ THE GAME IS OVER--  FINAL SCORES ARE: "  "COMPUTER= "Y "AND YOU HAVE " Z
180 PRINT "END OF GAME"
185 PRINT "IF YOU WANT TO START    OVER, PRESS RETURN KEY": INPUT G
190 GO TO 95
```

Program 8-6

Sec. 8-6   Electronic Safe Combination   117

loop commands and utilizes the random-number selection capability of the computer. The program enables the computer to keep a dual score and to print out one for you and one for the computer. Color symbols are utilized, although their inclusion is at the discretion of the operator. Similarly, graphic illustrations can be included to show the three shells, but this would materially add to the complexity of the program.

8-6. ELECTRONIC SAFE COMBINATION

In addition to the programs shown earlier, many others involving numbers as a central theme can be formulated. In some instances, however, a particular concept may involve a much longer program than those previously shown. A typical example is illustrated in Program 8-7, which challenges you to find the correct combination for a safe. Instead of a single dial for finding the combination, assume this safe has three numbered push buttons, and the combination must be entered in sequence by pressing these button switches on the door of the safe. Thus, the combination might be 3, 3, 2, or it might be a sequence such as 2, 1, 3. It is, of course, possible that the combination might be identical numerals such as 1, 1, 1 or 3, 3, 3, because the three numerals are individually selected at random by the computer. Once the combination has been selected by the computer it is not changed for that particular game. You get ten chances per game to find the correct combination. The computer keeps track of the number of turns taken and also notifies you when a correct numeral is selected during the game.

Some game strategy can be employed. For instance, if your first selection is 2, 1, 3 and the computer indicates that your first numeral is correct, you can use that same initial numeral and try to find the combination with new second and third numerals. If you use up ten turns and do not find the combination, the computer will notify you that you lost and the game is over. For computers containing timing provisions, line 195 indicates the time-check provision. If this is not desired, then lines 195 and 205 can be converted to PRINT statements.

The game, as shown in Program 8-7, provides a fairly interesting sequence, and it is not too difficult to win. The program can be expanded to a more difficult combination by specifying four electronic buttons. In the latter instance the random selection by the computer must extend from one through four. Four numerals, however, would involve many numerical combinations and the game would become very difficult to win.

```
30 GOSUB 860
35 PRINT "-THE ELECTRONIC SAFE-"
40 GOSUB 860
50 FOR T = 1 TO 5000: NEXT T
60 PRINT "▓ ♦♦♦♦♦   RULES   ♦♦♦♦ "
70 GOSUB 855
80 PRINT "▓THE SAFE HAS THREE     ELECTRONIC BUTTONS TO SELECT PROPER NUMERAL SEQ
UENCE"
90 PRINT
100 PRINT "▓NUMERALS ARE 1, 2, & 3 FOR OPENING SAFE"
105 PRINT
110 PRINT "▓THE SAFE COMBINATION  MAY BE 1,2,1 OR 3,3,2 AND SO ON"
120 GOSUB 860
125 PRINT "▓THE NUMERAL SEQUENCE  IS SELECTED BY THE     COMPUTER"
130 FOR T = 1 TO 4000:NEXT T
135 PRINT
140 PRINT "▓▓NUMERAL SEQUENCE IS   THE SAME THROUGHOUT   ONE GAME"
145 FOR T = 1 TO 3000:NEXT T
150 PRINT
155 A=INT (RND(1)*3)+1
160 B =INT (RND(1)*3)+1
165 C=INT(RND(1)*3)+1
170 N = 0
175 PRINT "▓YOU GET TEN CHANCES   PER GAME TO FIND THE   CORRECT COMBINATION"
180 PRINT
185 FOR T = 1 TO 2000:NEXT T
190 PRINT
195 PRINT "▓THE COMPUTER WILL BE  KEEPING A TIME CHECK"
200 PRINT
205 TI$ = "000000"
210 PRINT "▓NOW TYPE IN YOUR OWN  SEQUENCE CHOICE OF 1, 2, AND 3"
215 PRINT "▓SEPARATE EACH WITH A  COMMA AND PRESS THE    RETURN KEY":INPUT X,Y,Z
220 GOSUB 860
225 N = N+1
230 GOSUB 860
240 IF A = X THEN 630
250 IF N = 10 THEN 740
600 PRINT
610 PRINT "▓COMBINATION WAS NOT   FOUND--TRY AGAIN"
620 GO TO  210
630 PRINT " ▓▓ YOUR FIRST NUMBER" X " WAS CORRECT!"
635 IF B = Y THEN 655
640 IF N =  10 THEN 740
645 PRINT "▓▓YOUR SECOND NUMERAL  DID NOT MATCH--TRY    AGAIN"
650 GO TO 210
655 PRINT "YOUR SECOND NUMERAL " Y " WAS CORRECT!"
660 IF C = Z THEN 680
665 IF N = 10 THEN  740
670 PRINT "▓YOU MISSED OUT ON     YOUR THIRD NUMERAL--   TRY AGAIN"
675 GO TO 210
680 PRINT "▓▓YOUR THIRD NUMERAL " Z " WAS ALSO CORRECT    AND YOU OPENED THE    SA
FE ! "
685 PRINT
690 PRINT "▓▓THE COMBINATION OF    THE SAFE WAS:" A;B;C
695 PRINT
700 PRINT "▓▓ YOU ARE THE WINNER   "
705 PRINT
710 PRINT "YOU HAD A TOTAL OF " N " TURNS WITH TOTAL    TIME  OF  " TI$
713 FOR T = 1 TO 5000:NEXT T
715 PRINT
730 PRINT "▓ THE GAME IS OVER ! "
735 END
740 PRINT "▓  YOU HAVE USED UP     YOUR 10 TURNS AND     LOST THE GAME"
745 PRINT
750 PRINT "TO TRY AGAIN--WITH A   NEW COMBINATION-PRESS RUN AND RETURN KEYS"
755 END
850 PRINT
855 PRINT
860 PRINT
865 PRINT
870 RETURN
```

Program 8-7

SOLUTIONS TO CHAPTER OPENING MAZES

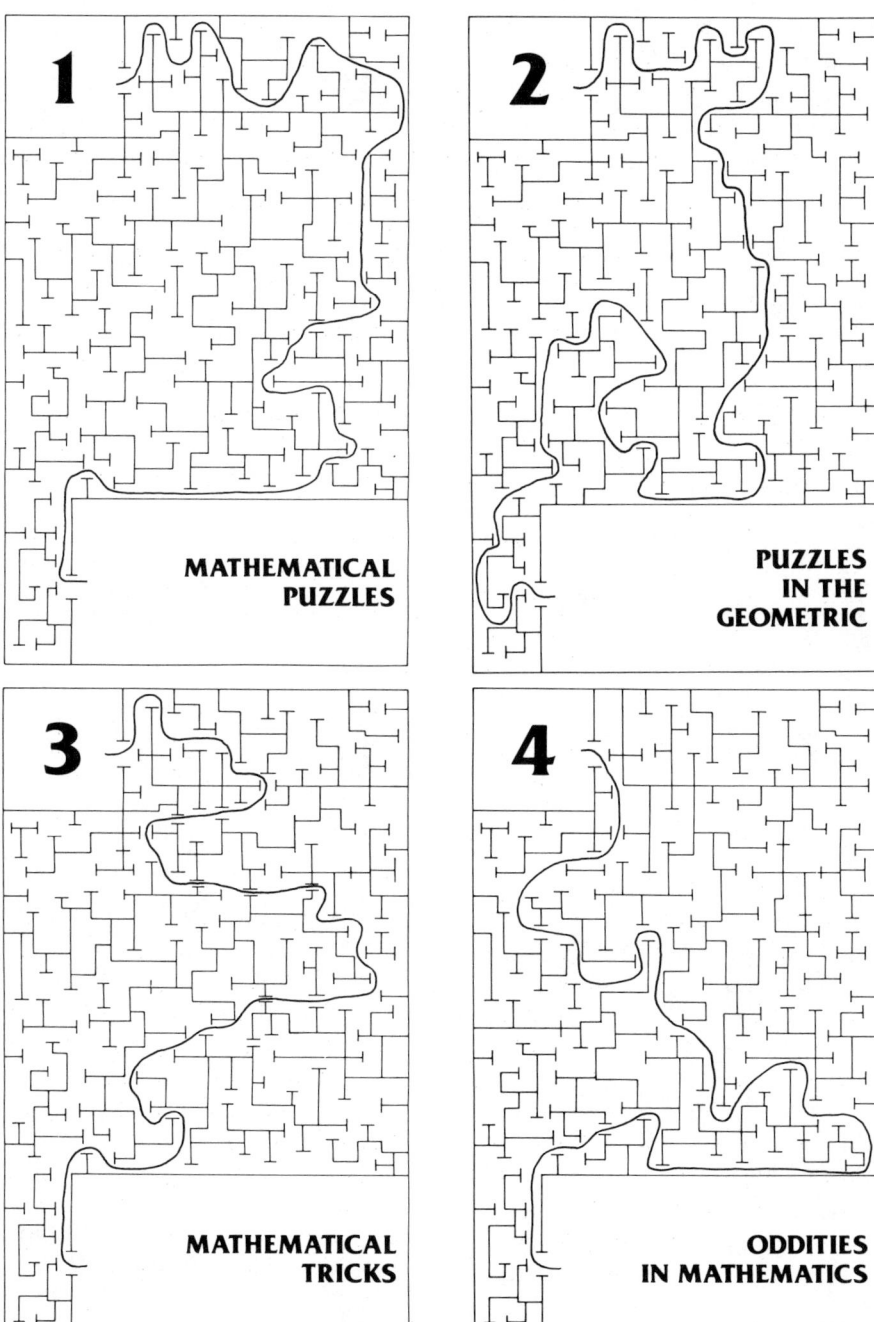

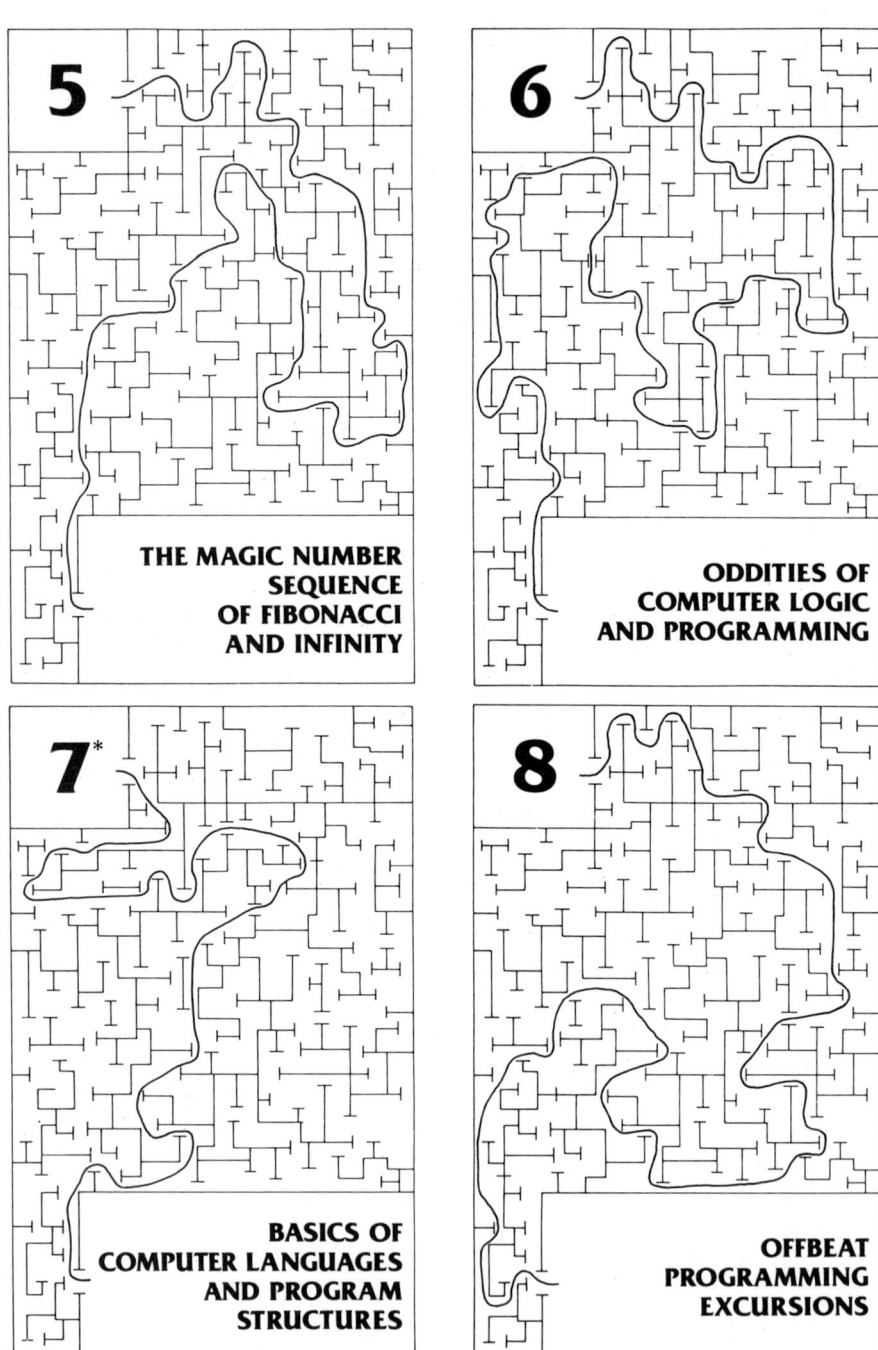

*Chapter 7 maze solution also solves title page maze.

INDEX

Algorithm, 101
APL, 99

Babbage, C., 100
Base-2 system, 73
BASIC, 100
Bigger piece of pi, 109
Binary system, 73, 74
   base-10 equivalents, 77
Boole, G., 74
   algebra, 89
Bountiful choice, 5
Brain twister, 7
Byron, A., 100

Calculators don't lie, 35
Cantor, G., 68
Carroll, L., 74
COBOL, 98
Complement, 80
Computer
   logic, 73
   programming, 83, 98
   system, 84
   trick, 111
Crunching, 112

Dialog Compromise, 82
Digit's sum equals the cube root, 52
Dodgson, C. L., 74
Double trouble, 5
Duplicates are repetitive, 54

Early-solution trick, 25
Electronic safe, 117

Factorial question, 5
Fast answer trick, 30
Fibonacci, 61
Flow chart, 101
Folding limit is 8, 54
For you it's easy, 36
FORTRAN, 83, 98

Golden ratio, 65

Hidden product, 32
How is it done?, 34

Infinity
   contradictions, 67
   It becomes its own 1/X, 55
   It proves nothing, 55
   It sounds complicated, 9
   It's almost impossible, 7

Let's look at quotients, 51
Letters in numbers, 8
Logic, 3
   system, 86
Logic and math, 73

Magic multiplier, 63
Magic number sequence, 61
Magic of 37037, 49
Mobius ring, 18
Mystery answer, 31
Mystic spirals, 64

Nines stand on their heads, 56
Numeral 8 is strange, 53
Numerical shell game, 33

Odd alternates, 49
Odd, strange, and complicated, 54
Oddities, 43, 73
Original number reappears, 44

Paper can't be stretched, 14
Paper cube blowout, 15
Pascal, B., 100
Peculiarities of repeated digits, 47
PL/1, 100
Pretzel caper, 20

Programming, 83, 98
  charting, 103
  structured, 101
Puzzles
  answers, 9
  bountiful choice, 5
  brain twister, 7
  double trouble, 5
  factorial question, 5
  it sounds complicated, 9
  it's almost impossible, 7
  letters in numbers, 8
  Mobius ring, 18
  paper can't be stretched, 14
  paper cube blowout, 15
  pretzel caper, 20
  reassemble and gain, 13
  relatively speaking, 5
  reverse logic, 3
  simple cube root, 6
  simple equation, 4
  the squares have it, 6
  trick question, 5, 9
  two-coin, 4
  upside-down messages, 8
  what pattern?, 7
  what's the cost?, 6
  what's wrong here, 5
  wheels within wheels, 19
  why always 26262626?, 6

Reassemble and gain, 13
Recurring sequence of ones, 43
Relatively speaking, 5
Repetitive reciprocals, 56
Reverse and subtract, 57
Reverse logic, 3
Round about turn-around, 36

Same answer repeated, 46
Secret number revealed, 28

Shell game, 116
Simple cube root, 6
Simple equation, 4
Strange oddity of nines, 44

The squares have it, 6
Thirteen-trick program, 113, 114
This one's easy, 35
Trick game is in your favor, 35
Trick question, 5, 9
Tricks
  calculators don't lie, 35
  early solution, 25
  fast answer, 30, 31
  for you it's easy, 36
  hidden product, 32
  how is it done?, 34
  mystery answer, 31
  numerical shell game, 33
  secret number revealed, 28
  this one's easy, 35
  trick game is in your favor, 35
  two tricky tricks, 37
Two-coin puzzle, 4
Two tricky tricks, 37

Upside-down messages, 8

Venn diagrams, 90

What pattern?, 7
What's the cost?, 6
What's wrong here?, 5
Wheels within wheels, 19
Why always 26262626?, 6
Why always 22332233?, 52